IEE Telecommunications Series 17
Series Editors: Professors J. E. Flood, C. J. Hughes and J. D. Parsons

Local
Telecommunications 2
into the digital era

Revised edition

Other volumes in this series

Local Telecommunications 2

into the digital era

Revised edition

Edited by J.M.Griffiths

Peter Peregrinus Ltd on behalf of the Institution of Electrical Engineers

Published by: Peter Peregrinus Ltd., London, United Kingdom

© 1986 Peter Peregrinus Ltd.

Reprinted with 3 additional chapters, 1988

While the author and the publishers believe that the information and guidance given in this work is correct, all parties must rely upon their own skill and judgment when making use of it. Neither the author nor the publishers assume any liability to anyone for any loss or damage caused by any error or omission in the work, whether such error or omission is the result of negligence or any other cause. Any and all such liability is disclaimed.

British Library Cataloguing in Publication Data

Local telecommunications.—(IEE
  telecommunications series; 17)
  2 : Into the digital era
  1. Telecommunication—Great Britain
  2. Telecommunication systems
  I. Griffiths, J. M. II. Series
  621.38′0941 TK5102.3.G7

ISBN 0-86341-080-4

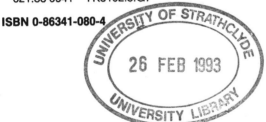
Printed in England by Short Run Press Ltd., Exeter

Contents

List of Contributors

Chapter 1
Eric Vogel
British Telecom Retired

Chapter 2
Bob Whorwood
Aston University

Chapters 3 and 18
Ron Brewster
Aston University

Chapter 4
Ian Paterson
CASE Communications Ltd

Chapter 5
Bob Whorwood
Aston University

Chapter 6
Gerry Lawrence
GEC Telecommunications Ltd

Chapter 7
Sydney F. Smith
STC Telecommunications Ltd

Chapter 8
Mark Trought
Plessey Network & Office Systems Ltd

Chapter 9
Peter Hughes
University College of North Wales

Chapter 10
Steve Wilbur
University College London

Chapter 11
Ken Eason and Leela Damodaran
Loughborough University of Technology

Chapter 12
Richard Boulter
British Telecom Research Centre

Chapter 13
Ken Quinton
British Cable Services

Chapter 14
Granville Taylor
British Telecom

Chapter 15
Bob Swain
British Telecom Research Centre

Chapter 16
Peter Studd
Ewbank Preece Telecommunications Ltd

Chapter 17
Ian Dufour
British Telecom

Chapter 19
David Fisher
STC Telecommunications Ltd

Chapter 20
John O'Reilly
University College of North Wales

Preface to new edition

Some say that Local Telecommunications requires a multidisciplinary
approach; others say it is just messy. This is certainly true, whichever
way you say it. Optical fibres, human factors, data protocols and video
standards are just a few items relevant to the local telecomms engineer.
The topic has changed rapidly. In 1972 when I first became involved
in the Local Network, everybody thought the step unwise and that I should
get a "proper" job in transmission or switching. In 1980 the IEE was
sceptical whether it was relevant to hold a Vacation School. By 1983 it
was felt sufficiently newsworthy that the Vacation School students notes
should be produced as a book. In 1986 it was clear that the original
book could do with updating and a second edition was issued. Now, in
1988 we have an extended version of that second edition.
The changes introduced in this book reflect the impact of the introduction of modern technology at a very high rate in the last few years.
Telephones and exchanges now realise the transmission functions by
semiconductor devices; mobile telecommunications has blossomed and the
Integrated Services Digital Network has become a reality.
This book represents the output of speakers at an IEE Vacation School
held at the University of Aston in the summer of 1986 with additional
papers from a further Vacation School held in 1988. It covers broadly
those aspects of local telecommunications which will be needed by an
engineer coming into this area for the first time, perhaps as a graduate
straight from University or with experience in a rather narrower area of
telecommunications such as terminal equipment, transmission or switching.
By the end of this book he or she will have been exposed to the other
techniques and technology needed to work in Local Telecommunications. The
area covered includes the local aspects of switching, signalling and
transmission for telephony and data purposes. Customer terminals are
also discussed both for their technical features and their human and
machine interfaces. Since the first edition of this book chapters have
also been added on the use of radio in the local network and for cordless
communication. Entertainment and communications by Cable TV networks are
also included, together with a look at how things are done away from the
United Kingdom. A chapter specifically related to the ISDN has been
added. In 1988 further chapters were added on optical fibres and the ISO
seven-layer model.
Obviously in a book of this size and with the breadth of subjects
covered, topics can only be tackled in a fairly superficial manner.
However, it is a lead-in to other, more detailed specialist works.
I must thank all the contributors for the time and effort they have
put into their contributions, and Sally Cable for helping me to put it
all together.

John Griffiths
May 1988

Chapter 1

The present local network

Eric Vogel

1.1 INTRODUCTION

For the purpose of this book we are considering the local network to include that part of the telecoms network from the customer to the local exchange. In the past the term "local network" has sometimes been taken to include the "Junctions" linking the local exchanges but this book does not include that area.

At one time the boundary between the local network and the exchange was well defined. However, the introduction of small low cost switching systems connected within the local network as a means of economising in plant or to overcome short term pair shortages has effectively eroded this once rigid demarcation between a telephone exchange and the local network. Furthermore, the possible integration of cable television and other services with the local network may introduce TV receiving and distribution nodes in the network, while business premises can become terminals of large digital systems in their own right.

Therefore, nowadays and increasingly in the future, a clear demarcation point of the local network is difficult to define. It is convenient to consider the local network as a medium for connecting a whole range of services to subscribers' premises, but clearly recognising that its existing configuration is one that has evolved essentially to connect to the local exchange a very large number of subscribers for telephone (audio) purposes. It is from that form that its further evolution and exploitation must be considered.

1.2 THE COMPOSITION OF THE UNITED KINGDOM LOCAL NETWORK

The British Telecom local network comprises cables, ducts, manholes, jointing chambers, flexibility cabinets and poles that serve to connect some 20 million subscribers to their local exchange. A simplified presentation of the arrangements is illustrated in Fig 1.1

It is a radiating arrangement divided into two separate parts:

(a)   The main network between exchanges and flexibility cabinets - generally referred to as the 'E' side.

(b)   The distribution network between flexibility cabinets and distribution points, most commonly poles, from which subscribers are finally connected - generally referred to as the 'D' side.

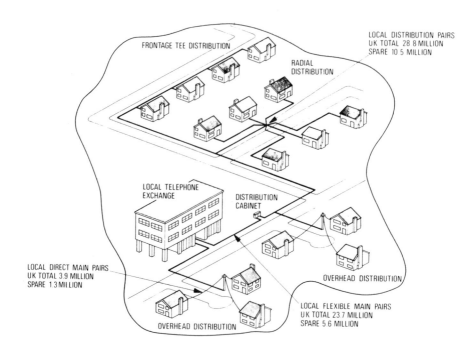

Fig 1.1 ARRANGEMENTS OF BRITISH TELECOM LOCAL NETWORK

The main network consists of relatively large multi-pair cables, ranging in size from 100 to 4800 pairs, laid underground in duct to facilitate renewal or enhancement without the need to dig up streets.

For subscribers close to the exchange, cables are laid directly to the distribution points because the route length is so short that no advantage arises from interposing a flexibility cabinet. Also in some rural areas (especially those with scattered penetration) flexibility cabinets are not installed although a cabinet is sometimes provided to give maintenance access.

The distribution network, in contrast, consists of smaller cables ranging in size from 2 to 100 pairs. The majority of these cables are also laid in underground duct but, for a period, directly buried cables were used especially on new housing estates. Directly buried cables have proved to be a maintenance hazard due mainly to other utilities working parties and the practice has been discontinued.

The objective of dividing the network into main and distribution parts by flexibility cabinets is illustrated in Fig 1.2.

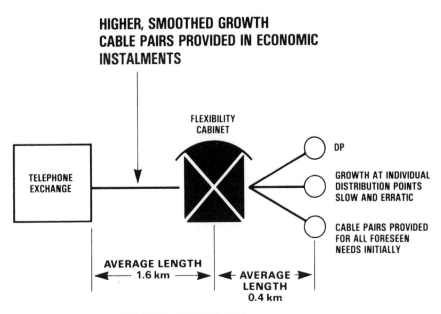

**HIGHER, SMOOTHED GROWTH
CABLE PAIRS PROVIDED IN ECONOMIC
INSTALMENTS**

FLEXIBILITY
CABINET

TELEPHONE
EXCHANGE

DP

GROWTH AT INDIVIDUAL
DISTRIBUTION POINTS
SLOW AND ERRATIC

CABLE PAIRS PROVIDED
FOR ALL FORESEEN
NEEDS INITIALLY

AVERAGE LENGTH
◄— 1.6 km —►

◄ AVERAGE ►
LENGTH
0.4 km

Fig 1.2 THE PRINCIPLE OF FLEXIBILITY

The distribution points serve limited numbers of subscribers; thus the growth rate is low, erratic and difficult to forecast. Because the growth rate in the main parts of the network are the summation of the growth rates of the distribution parts, they are higher, less erratic and more predictable. Resulting from this, the main network cables can be provided in economic installments and connected to the distribution network cables via the flexibility cabinet facilities.

For the relative short route lengths of the distribution network, cables are installed at the outset with sufficient pairs to meet the ultimate requirements. The distribution cables are generally so small that deferment of provision of pairs is uneconomic. The greatest advantage of initially providing for the ultimate requirements is that however erratic the take up may be there is always sufficient pairs to meet demand for service.

Flexibility is a key factor to achieve economic installation of a local network. British Telecom is a leading proponent of the technique, having introduced flexibility cabinets during 1946. Elsewhere, notably in America and Japan, a different method has been used for providing flexibility whereby pairs are 'teed' at one or more points in the cable route and are thus available at different locations. This method is now generally judged as unsatisfactory because it complicates the updating of records, degrades the transmission parameters and above all prevents the satisfactory operation of digital transmission by reason of multiple echos arising from unterminated 'tees'. Flexibility cabinets also have the advantage that they provide an access point for testing and fault location.

1.3   PLANT HARDWARE

1.3.1   DUCT

The duct network in which cables are run is very extensive. It radiates from the telephone exchanges to reach within a few tens of metres of most houses in the UK; it usually extends right into major business premises.

The ducts are made of PVC, glazed earthenware, or steel. Those made of earthenware, which are of 92mm bore-diameter, are laid in 1 metre lengths; PVC duct of 51mm or 90mm and steel duct with 102mm bore-diameter, are usually laid in 6-metre lengths. Lengths are jointed together by spigot and socket connexions or by sleeved joints and they provide a reasonably tight fit, but for economic reasons no attempt is made to make the duct bores watertight or gas-tight.

Ducts may be laid singly or in multiples up to 96 ways; those formations above 9-ways are provided with concrete supporting walls. Ducts laid in the highway normally have at least 600mm, and in the footway about 350mm, of cover. Steel duct is mainly used in situations where a shallow depth of cover is only possible.

1.3.2   JOINTING CHAMBERS

Surface access joint-boxes or underground manholes are provided at intervals of about 200m along the duct route, providing access to the plant for the installation of new plant, and the re-arrangement and maintenance of existing plant.

The jointing chambers are constructed in reinforced concrete or brick, although some smaller boxes may be precast (eg in a glass resin fibre). Brackets and bearers carry cables with enough space available for making and supporting cable joints. Duct routes and manholes may be used for junction and trunk as well as local cables; manholes in these cases also provide convenient locations for loading pots, repeater cases, etc.

The size of joint boxes and manholes is determined by the number of ducts likely to be required and the size of cable joints to be constructed and accommodated.   The strength of walls, roofs and access covers are designed to withstand the combined forces from earth pressures and traffic loading.

1.3.3   LOCAL LINE CABLES

Cables in the main network between the exchange and its cabinets have aluminium alloy or copper conductors with cellular polyethylene insulation and polyethylene sheaths. The sheaths incorporate a 150mm aluminium foil barrier fused to the inner surface of the polyethylene sheath to prevent the ingress of moisture by molecular migration.

Cables with copper conductors use gauges of 0.32, 0.4, 0.5 and 0.63mm diameter, the largest cable size being 4800 pairs; aluminium alloy of 0.5mm gauge is used in cable sizes up to 2000 pairs. All these cables are installed in duct.

To protect main cables against loss of insulation resistance or corrosion of conductors due to ingress of moisture at faulty joints or through cable sheath damage, resin air blocks are introduced at the exchange and cabinet ends of each cable system and air at a pressure of 620mbar is injected into the cables from the exchange. The air is supplied via a desiccator to make sure no moisture is introduced. Current plant and construction methods should make it possible for all new cables to meet minimum pressure standards of 500mbar for distances 0-2km from the exchange and 450mbar beyond 2km. The desiccated air flow is continuously monitored by flow meters and pressure gauges in the exchange. Gauges installed in cabinets and contactors attached to the cable operate an electrical alarm at the exchange over an alarm circuit if the pressure falls to a pre-determined level.

Distribution cables connect distribution points to cabinets and are up to 100 pair in size. They are made up of 0.5 and 0.6mm copper conductors or 0.5mm aluminium conductors with cellular polyethylene insulation. Unlike the air-filled main cables the cable cores are filled with petroleum jelly to prevent the ingress and lateral flow of water. This type of cable is for the use in duct. A similar cable but protected against mechanical damage by a layer of galvanised wire-armour for direct burial in the ground has been used but as mentioned earlier this method is discontinued. A figure-of-eight cable with a simple polyethylene sheath enclosing a cable core in one part and a steel suspension wire in the other is used as a self-supporting aerial cable for erection on poles.

1.3.4 WIRE JOINTS

The wires of cables, identified by colour coded insulations and made up into units of cable pairs, are jointed by insulated wire connectors, crimped by hand tools or hydraulic powered jointing machines. Where a number of consecutive joints are to be made in a main cable, the pairs in a 100-unit group may be jointed at random to simplify the jointing technique and later identified and marked at the terminal point using plug-in automatic pair identification equipment at the exchange.

1.3.5 JOINT CLOSURES

A number of methods have been used in an effort to obtain an efficient closure of polyethylene sheathed cable comparable with the standard of reliability achieved with plumbed sleeves on lead sheathed cables.

The main methods now in use are:-

(a)   "Shrink-down". This method utilizes a cross-linked polyethylene sleeve impregnated with a sealing compound. When heat is applied to the sleeve it contracts into a tightly fitted form enclosing the jointed wires.

(b)   Injection Welded. Moulds are fitted over the end of the polyethylene jointing sleeve and the cable, and polyethylene is injected under a temperature-controlled process to seal the joint by bonding the parts together.

(c)   Polyethylene Sleeve Mechanical Joint.   This type of closure is used mainly in the distribution part of the network.   The cables entering the joint are resin-sealed to the base, while the top of the joint is closed by a cap-end held in position by a hinged clamp. Increasing use of this closure is being made for housing equipment of new systems now being installed.

These methods have now been adopted in place of a variety of earlier methods that all proved unreliable.   The injection-welded joint is proving to be most reliable although the equipment needed to make the closure is both bulky and expensive.

1.3.6   FLEXIBILITY CABINETS

The inter-connection of the pairs in main and distribution cables is effected in cast iron cabinets, which are made in two sizes to accommodate either four or eight cross-connexion strips.   The wires to be connected pass through identification holes in the strips and the wire connections are made with the grease-filled crimped connectors.

1.3.7   OVERHEAD PLANT

Poles used for outside plant vary in size according to the situation but most are around 11 metres in length.   They are usually of Scots Pine and are creosoted by a pressurised process.   Routine tests for decay in every wood pole are made every 6 years to ensure safety for the public and staff.

Hollow poles made of reinforced plastic or stainless steel are now available as alternatives to the traditional wooden pole.   These types of poles facilitate the installation of the self-supporting aerial cable to be made without the need to climb the pole.   Although more costly to manufacture, they have the advantage of not requiring routine tests for decay and are safer for staff to work on.

Poles remain an efficient and economical way of distributing wires and developments in this field are continuing.   Once, pole erection was very labour intensive but now poles are increasingly taken to site and erected by a Pole Erection vehicle which has a number of hydraulic aids.   The pole is loaded by the vehicle boom, and later lifted from the vehicle into a hole which has been bored by an earth auger mounted on the boom.

1.4   DISTRIBUTION METHODS

Distribution points, served by local line cables to provide service to subscribers fall mainly into one of the following categories.

(a)   Internal - a terminal box or frame within a building.

(b)   External - an external wall-mounted terminal block in the case of small buildings or shops.

(c)   Underground - the mechanical access cap-ended joint closure housed in a joint box from which the subscribers service cables radiate.

(d)   Jointing Post - a similar interconnection of subscribers service cables to above but housed above ground in a galvanised steel channel with removable cover, 0.5 metres high, located against a wall.

(e)   Overhead - a terminal block mounted at the top of a wood pole from which the self-supporting aerial cables radiate.

On new housing estates, dwellings are served in the usual way at no cost to the developer using the cheapest method considered to be satisfactory by BT. This will usually be by the overhead method of distribution for low-rise dwellings and internal terminal boxes for high rise flats. If, however, a more expensive system is required by the estate developer, the method used is negotiated with the developer by the planning staff.

1.5 QUANTITIES AND BOOK VALUE OF PLANT

To give an indication of the size of British Telecoms local network. Fig 1.3 shows approximate quantities of plant as of 1985.

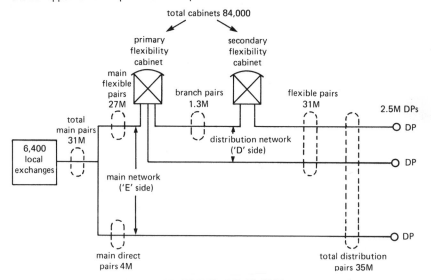

Fig 1.3 APPROXIMATE QUANTITIES OF PLANT

The total number of pairs account for some 48 million pair km in 495 thousand sheath km of cable and 124 thousand route km of duct. In addition some 4 million poles are used.

This large quantity of plant amounts to a total book value of some £1500M with an annual expenditure in adding to and renewing the plant approaching £200M.

In terms of the overall investment for the whole telephone system of the UK, the local network accounts for some 22% and is only exceeded by the value of exchange equipment at 39%.

This large scale of investment, and recognising that for the immediate future the same type of plant will in the main be installed, emphasises a need to investigate all possible ways of exploiting the local network.

1.6 PLANNING OBJECTIVES AND PRINCIPLES

The objective in planning the installation of local network plant is to ensure that cable pairs are readily available to meet demands wherever and whenever they occur.

The obvious solution is to provide ample pairs at all points of the cable network for all foreseeable demands. This solution results in poor utilisation (ie the ratio of working pairs to total pairs) as well as introducing a financial burden.

In practise there are some significant economics of scale possible and the economic policy is to install plant in relative large increments in advance.

For duct and for the distribution cables at the fringe of the cable network the fixed costs of installing the plant is so high, compared with the marginal cost per added unit, that once-and-for-all installation to meet the ultimate foreseeable demand is normally justified.

For main cables, as mentioned earlier, incremental installation is economic because the use of duct reduces the cost of installing cables and because flexibility cabinets tend to smooth and concentrate growth onto a limited number of cable routes.

In order to install cable pairs in advance of take-up, and avoid a financial burden, it is necessary to monitor pairs take-up at every flexibility points. This is achieved by maintaining records such as that shown in Fig 1.4.

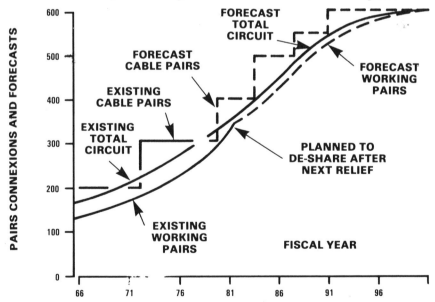

Fig 1.4 GRAPH OF TYPICAL FLEXIBILITY POINT

By comparing the number of pairs available with those forecast to be required, the time which provision of new pairs must be initiated can be estimated.

Forecasts are prepared by the marketing department staff who carry out surveys, talk to local councils regarding proposed developments and determine the likely requirements of business customers. Thus they build up a demand forecast for each flexibility cabinet area for a period of 18 to 23 years depending upon when the forecast is prepared. The forecasts are given for each year for the first five years and then at intervals which align with the National population census and inter-census years (every five years). The forecast demand at each date is expressed in terms of residential, business and miscellaneous; the miscellaneous includes private wires, telex circuits etc. Computer programs are available to planning engineers which assist them in determining cable pair utilisation and' the design at cable relief schemes. By following the principles British Telecom achieve a utilisation factor of some 70% for the main network and 65% for the distribution network.

1.7 PLANNING OPTIONS

The present method of connecting subscribers to the local exchange by individual pairs of wires is usually the cheapest and simplest method in UK. Because of the comparatively high end-equipment costs of multiplexing systems and the relatively short lengths of line, possible line plant savings do not approach additional equipment costs. In countries where the length of lines are much longer (eg America) the use of multiplexing systems become more economical and are extensively used.

Limited use is, however, made of a simple 1+1 subscribers carrier system providing two completely independent telephone circuits on one pair of wires. Although it is not an economic alternative to the provision of the full number of pairs at the outset, it is valuable as an expedient to provide service at short notice where there is a shortage of pairs. These systems were extensively used in the 1970's to reduce the waiting list.

Similarly, use is made of remote line concentrators when it is not possible to provide additional main cable pairs immediately. Two versions are available:-

(a)   A roadside unit normally installed alongside a flexibility cabinet that allows 96 subscribers to be served by 16 main cable pairs.

(b)   A wall or pole mounted unit that allows 14 subscribers to be served by 5 main cable pairs.

Whilst the planning objectives, principles and options are highly important to achieve an economic installation of local line plant, equally important are the choice of conductor sizes to economically meet the requirements for the signalling and transmission parameters.

1.8 SIGNALLING AND TRANSMISSION PARAMETERS

The resistance of local pairs is restricted to limits which ensures the satisfactory interworking of telephones and local exchanges and varies with type of exchange. The limit for electro-mechanical exchanges which still dominate in the UK is 1250 ohms. Where standard limits are exceeded by normal cable installation an electronic line-current boosting device is included to extend the limit to 2000 ohms.

The attenuation limit for local pairs is set by the overall network transmission plan which is designed to ensure satisfactory transmission performance for national and international calls. Fig 1.5 shows the transmission plan which shows that the local pair must not exceed 10dB (at 1600 Hz).

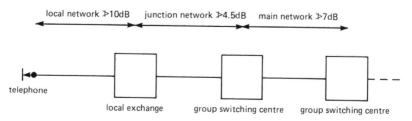

Fig 1.5   ATTENUATION LIMITS FOR THE OVERALL NETWORK AT 1600 Hz

The 10dB limit is seldom approached because line lengths in the UK are relatively short and the range of conductor sizes, as mentioned in paragraph 1.3.3, enable the routes to be planned below the limit. In the minority of cases where the limit is exceeded a fixed gain amplifier is fitted.

The electronic devices to enhance signalling and transmission performance are installed at the local exchange and are known as Line Extenders. Detailed information about these devices are given in Chapter 14, Electronics in the Local Network.

1.9 LOCATION AND SAFETY OF LOCAL LINE PLANT

In parallel with meeting the above mentioned planning and performance requirements, it is important to recognise the location limitations and safety aspects of local line networks.

The ducts, joint-boxes and manholes used in the local line network have to share the limited space under streets with plant owned by other utilities such as Electricity, Water, Gas, Sewer and Underground Railway Authorities. Careful planning is necessary and difficult construction problems are often encountered, especially in congested city centres where telephone duct systems are large and congestion from other utilities plant arise.

Preferred locations for the different underground services have been agreed by the National Joint Utilities Group and utilities do their best to conform to these recommendations. However, plant congestion and the effect of street alterations often make this impossible and each utility is obliged to install its plant in the best way possible.

Under such conditions, not surprisingly, serious damage is often caused to telphone outside plant by the operations of other utilities - and vice-versa. Incidents of this sort are a significant cause of interuption to telephone service and extensive arrangements are made to oversee such operations in progress.

Gas leaking from faulty mains is likely to enter telephone ducts and is a particular source of danger to staff working in manholes and in buildings connected to the duct system. Duct entries into all exchanges, customers premises, cabinets or kiosks are therefore carefully sealed, and staff working in all underground structures are required to make regular tests with gas detectors.

1.10 UTILISATION OF THE LOCAL NETWORK

The local network is overwhelmingly used for telephony exchange lines and Fig 1.6 shows how small a part (less than 11%) of the plant is utilised for miscellaneous circuits. Apart from some external extensions and private circuits, all exchange lines are connected directly via the Public Switched Telephone Network (PSTN).

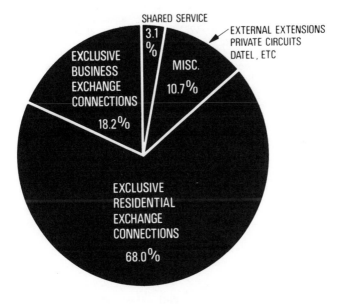

Fig 1.6 UTILISATION OF PAIRS IN THE UK LOCAL NETWORK

In terms of overall utilisation two other factors need to be considered:

(a)   Pair utilisation as discussed in Paragraph 1.6

(b)   The amount of traffic carried by pairs.

The current levels of pair utilisation, 70% for the main network and 65% for the distribution network are probably about optimum as a compromise between meeting demands for service and keeping asset liability to an acceptable degree.

National statistics for the UK show that the average useage of a telephone is less than 2% of the total available time. This means that for over 98% of the available time the average local line is idle.

This extremely low traffic utilisation has long been recognised and various features and services, both telephony and other, have been added to stimulate traffic. Plenty more scope remains and bearing in mind the very large asset value of the local line plant mentioned in Paragraph 1.5 there is an overwhelming need to further exploit the local network.

1.11   EXPLOITATION OF THE LOCAL NETWORK FOR DIGITAL WORKING

1.11.1   GENERAL

For almost two decades British Telecom have been installing and operating digital systems (Pulse Code Modulation Systems) in the junction network. More recently high speed digital transmission systems (120 and 140 Mbit/s) have been installed in the main (trunk) network using both coaxial and optical fibre cables and in 1981 the first local and main "all digital" (System X) exchanges were brought into service.

Within the next two decades it is forseen that the whole of the junction, trunk networks and all exchanges will be digital.

To take full advantage of this digital era and to meet the need to exploit the existing local network, a digital solution is required so that digital transmission facilities are available to any subscriber.

Later chapters will discuss in detail the various options for digital transmission and the range of services potentially available. It will suffice in the following paragraphs to review the salient technical considerations that arise when digital exploitation of the existing local network is contemplated.

1.11.2   TECHNICAL CONSIDERATIONS

The earlier paragraphs of this Chapter clearly indicate that the local line network is planned and installed as a medium for the transmission of analogue telephony (audio).

However, as local lines have no defined cut-off frequency they can also carry high frequency signals, limited only by attenuation, noise and crosstalk. Indeed, as Figure 1.7 shows, a large number of services can be carried, without special provision, on the local network; a notable exception being the transmission of high quality moving picutres.

Nevertheless, in terms of possible cable fill (the number of useable pairs), the insertion loss, crosstalk and noise aspects set an upper limit of about 100 kHz to high frequency systems in the network.

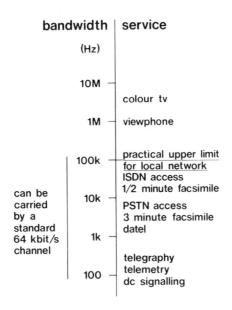

bandwidth	service
(Hz)	
10M —	
	colour tv
1M —	viewphone
100k —	practical upper limit for local network
	ISDN access
	1/2 minute facsimile
10k —	PSTN access
	3 minute facsimile
	datel
1k —	
	telegraphy
100 —	telemetry
	dc signalling

(can be carried by a standard 64 kbit/s channel — bracketing 1k to 100k range)

Fig 1.7   BANDWIDTH OF SERVICES

Typical values for 2km of 0.4mm copper conductor cables are shown in Table 1.1.

INSERTION LOSS/CROSSTALK	FREQUENCY kHz	
	100	400
Insertion Loss	20dB	40dB
Near-End Crosstalk.Mean Std. Dev.	80dB	71dB
Fear-End-Signal-To- Mean	75dB	63dB
Crosstalk Ratio Std. Dev.	9.5dB	9.5dB

TABLE 1.1   INSERTION LOSS/CROSSTALK VALUES OF 2km, 0.4mm CABLE

Insertion loss (expressed in decibels) increases approximately with the square root of frequency and, within any cable, its value is consistent within small limits of variation.   It is also proportional to length for similar types of cable.

There are two manifestations of crosstalk interference of significance to the system designer and operator - near-end crosstalk and far-end crosstalk.

Far-end crosstalk may be defined as an unwanted signal arriving at the distant end of a cable pair, arising from stray coupling from a neighbouring pair(s) transmitting in the same direction.

Near-end crosstalk may be defined as an unwanted signal received at the near end of a cable pair, arising from stray coupling from a neighbouring pair(s) transmitting in the opposite direction.

Far-end crosstalk, defined in terms of the far-end-signal-to-crosstalk ratio (FESXTR) when insertion loss is taken into account, tends to fall in a fairly regular manner with increasing frequency at 6dB/Octave.

However, at a critical frequency, around 2 MHz for routes of about 2km, and higher for shorter routes, FESXTR rapidly worsens.

Near-end crosstalk, defined in terms of near-end crosstalk attenuation (NEXTA) tends to be irregular over particular sections of the frequency spectrum with large narrow peaks of higher attenuation superimposed on the characteristic fall of 4.5dB/Octave.

There is little correlation between the physical features of cable routes and their crosstalk characteristics. Long or short routes are equally likely to produce 'good' or 'bad' results. This arises because any cable route is a mixture of large and small cable sizes and conductor sizes.

In general the majority of crosstalk interference is contributed via pairs in close proximity to disturbed pair because these pairs tend to have worse crosstalk values.

Due to the extreme variability of cable parameters it has proved to be an unrealistic task to model the parameters of the local cable network. Nevertheless, in terms of insertion loss and crosstalk there is ample evidence that the local network can support a useful and acceptable number of digital systems.

Impulsive noise and radio interference are other mechanisms that cause disturbance to systems.

Impulsive noise is induced into local cables from a number of sources:-

1)   other pairs in the cable connected to electromechanical telephone exchange equipment which produces hostile conditions, particularly during periods of dialling, ringing and switch hook operation.

2)   switching transients on power lines.

3)   electric traction systems (eg railways).

The disturbance is generally of large amplitude, short duration bursts, interspersed by various periods of no activity.

The problem with the assessment of impulsive noise is the difficulty of correlating the characteristic features of the interference with the impairment created, there being little consistency in its effect on different services.

The effect on audio speech circuits is to produce "clicks" in the receiver which, unless the disturbance is persistent or periodic does not prove too unacceptable to the listener. On a digital circuit impulsive noise will, of course, increase the error rate, and the higher the transmission speed the greater could be the degree of corruption caused to the signal. The error performance objectives for one local end of a 64 kbit/s Hypothetical Reference Connection are expected to be along the lines that 99% of seconds shall be error-free and 99% of minutes shall have a bit error rate of better that 1 in 10^6.

To what extent impulsive noise will frustrate these objectives and thus become a limiting factor in the digital exploitation of the local network has yet to be fully determined but evidence to date is highly encouraging.

Radio interference in general is not a problem. The fact that most cables are laid underground, together with their internal aluminium foil barrier very effectively screens the cable pairs from radio interference. Overhead sections of local pairs can be affected especially when sited near high powered radio transmitters but problems have only arisen on long sections of overhead pairs where 1+1 subscribers carrier are fitted. Because digital systems are inherently more robust to interference compared to analogue systems, radio interference should not be a difficulty when the local network is exploited for digital working.

Although the existing local cable network is not an ideal medium for the transmission of digital signals, it is expected that with the development of new digital techniques that limit the pertinent frequencies to be lower than 100 kHz, a useful and cost effective exploitation is possible.

1.12 BROADBAND TRANSMISSION

1.12.1 COMMUNITY ANTENNA TV NETWORKS

At present broadband transmission is a minor part of the local network scene.

British Telecom have been involved since 1969 in their installation especially for the provision of Community Antenna TV networks (CATV) but these until more recently have been restricted by legislation to some new town developments (eg Milton Keynes). Five such networks have been installed and concern the basic distribution of the four national TV channels. Where legislation permits, distant channels are 'imported' and carried by the networks to attract subscribers. All told some 50 thousand subscribers are served by these British Telecom networks.

The topology used for these networks is the coaxial cable 'tree' structure without backward path facilities, although provision is made in the design to cater for this in the trunk parts of the networks. The networks are overlays to the local cable network.

In 1983 the Government initiated a plan to 'recable Britain', the initial phase evolving from recommendations made by the Hunt Report.   Apart from stimulating a new general interest in broadband networks, one of the main thrusts was to initiate the provision of networks capable of meeting the demands of a predicted information technology explosion ie the two-way transfer of date information (interactivity) between subscribers and nodes to facilitate services such as electronic banking and teleshopping. Also permitted is the provision of 'premium' TV services for which subscribers pay additional subscriptions.

For this initial phase 11 franchises were awarded to a number of network providers to install networks in nominated cities, again as overlays to the local cable network.

Where considering the planning of wideband systems for CATV, two basic topologies are available:

(a)   tree structure

(b)   switched star structure

Details of these structures are given in Chapter 13.

In the event of wide application being viable the question arises, whether to integrate wideband and telephony services to form an integrated network?   Studies have shown that expanding the existing local network using simple pairs of wires to achieve 100% penetration is cheaper than providing an integrated wideband network.

Hence it would appear that for the foreseeable future the existing local network will continue to expand in its own right with separate wideband networks being installed where commercially justified.

1.12.2   DIGITAL WIDEBAND TRANSMISSION

At the beginning of this Chapter mention is made of business premises becoming terminals of large digital systems in their own right.

Several systems falling into this category have been installed by British Telecom.   The digital link between the local exchange and the business premises is provided by digital line systems operating at 2.048 Mbit/s using specially laid cables.   The pairs in these cables are separated into two halves by a transverse metallic screen, the respective transmission directions of the pairs being allocated to the separate halves.   The screen mitigates near-end crosstalk and enables a 100% cable fill.   The route lengths of these links are generally short and do not require intermediate regenerators but if necessary regenerators can be installed using standard digital junction circuit equipment.

In the immediate future optical fibre cables are planned to be used as an alternative to these special cables. Over the last few years the cost of optical fibre transmission systems have significantly reduced and are cost attractive for application to the local network scene. Cables ranging in size from 5 to 30 fibres are available and can be used for digital rates at 2.048, 8448 or 140 Mbit/s, according to the traffic requirements.

Optical fibre cables have many advantages over conventional types of cables. They are small and light weight enabling long lengths of cabling to be installed, cater for various digital rates without any need for intermediate regenerators and are unaffected by the ingress of moisture should cable sheath or joint closures fail.

It is forseen that optical fibre cables will become the standard form of provision in the local network to meet the needs of digital wideband services.

1.13 SUMMARY

The basic arrangements for installing local networks using dedicated pairs from exchanges to subscribers has virtually remained unchanged since their inception.

It is expected for the next decade or so that local networks will continue to grow towards 100% penetration, sustaining the same basic arrangements. New technology will enable a significant exploitation to provide a whole range of new services.

In parallel with this growth additional overlay networks will be installed to meet various broadband transmission requirements and perhaps it will be well into the next century before a fully integrated wideband transmission network is justified.

1.14 ACKNOWLEDGEMENT

I would like to thank the many people in British Telecom who have contributed to the compilation of this chapter.

1.15 BIBLIOGRAPHY

CCITT MANUAL. Local Network Planning - GENEVA 1979

CLOW, D G and STENSON, D W   External Plant - Post Office Electrical Engineers Journal, VOL.74 PART 3. OCT 1981

NATIONAL JOINT UTILITIES GROUP - Provision of Mains and Services by Public Utilities on Residential Estates - NJUG Publication NO 2 NOV 1979

VOGEL, F C and TAYLOR, C G BRITISH Telecom's Experience of Digital Transmission in the Local Network - PROC. ISSLS 82

Chapter 2

Voice communication requirements

Bob Whorwood

2.1 INTRODUCTION

For practically the whole of its history, the public telecommunications network has been designed primarily for voice communication. Speech traffic has earned the revenue necessary to establish and maintain the network, and thus made possible the provision of other services not economically viable in isolation. Non-speech services have been limited by the need to give priority to the requirements of speech transmission and its ability to earn the bulk of the revenue.

In recent years this situation has been rapidly changing because of the widespread introduction of digital technology. Speech traffic remains dominant, but meeting its requirements does not inhibit the provision of non-speech services in the way it is inhibited in an analogue environment. While an incentive to adopt digital technology is the opportunity it gives to provide an ever increasing range of non-speech services, the main incentive is the cheaper provision of improved speech services. After evolution over more than 100 years, it would be surprising if the basic requirements for a speech network were not already very well, if not fully, met. Digital technology enables these requirements to be met much more easily, as just one, albeit the dominant one, of the wide range of communication facilities becoming commonplace on the public network.

Although the integrated services digital network (ISDN) will ultimately carry all services involving speech, data, text, pictures etc., on the same bearers, they may be conceived of as having almost, if not quite, an independent existence. Thus there is less need to degrade one service to benefit another. The digital network presents to the telephone engineer both the necessity, and more importantly the opportunity, to re-examine the problems of speech transmission. Many of the constraints of the past are removed, but new problems are introduced. New options are made available, typified by the change from assuming a nominal bandwidth of 3.1kHz (300-3400Hz) to thinking in terms of a 64kbits/s channel.

Options arise concerning the best use that can be made of the available 64kbits/s, such as adaptive differential PCM requiring less capacity, thus freeing some bits for other, possibly non-speech application. Interest is perhaps less in what has already been decided upon, than in the greater flexibility available for future options. There can be little doubt that new services will be offered in the future which, at present, have not been conceived of and digital technology will ease that process.

2.1.1 Introduction to pulse code modulation (PCM)

In a limited way digital tranmission has been deployed in the telephone network for some years. Pulse Code Modulation systems comprising 24 or 3Ø channels have served as elements within an otherwise analogue environment. For the most part such systems have been used inside city conurbations and a few kilometres only in length. Conversion to digital form of transmission and back to analogue takes place at each end of such short systems and in no real sense affects the overall analogue nature of the network. Such systems merely enable greater use to be made of existing cables and the motive for their use is the avoidance of the need to lay extra cables to meet growth in demand for service.

Analogue systems, as the name implies, employ electrical signals which vary directly in sympathy with the acoustical signals they represent. Deviations from a precise one-to-one relationship between acoustical and electrical signals are regarded as distortion.

Digital transmission of electrical analogue signals is possible because the amplitude of samples, providing there is a sufficient number of them and they are taken at certain discrete instants in time, completely defines the waveform. As this is so, it follows that no information is contained in the waveform between these instants which must occur at a rate at least equal to twice the highest frequency component present in a complex waveform such as that of speech.

However, to achieve adequate transmission quality it is not necessary that the sample values be measured exactly, and a set of discrete values are chosen to represent the whole amplitude range of the analogue signal. This is known as quantizing, and the difference between the true amplitude and the discrete value chosen to represent it is known as quantizing distortion. The amount of quantizing distortion that can be tolerated depends upon subjective effects and, if it is imperceptible, there is clearly no need to reduce it. The amount of quantizing error that is acceptable depends upon signal amplitude, and the greatest accuracy required when the signal is small. Unnecessary precision is wasteful, and the required accuracy is related to signal size by logarithmic relationships known as companding laws.

The sample amplitude values may be transmitted and the

waveform restored at the far end, this being known as Pulse Amplitude Modulation (PAM). In the intervals between samples being sent, samples of other waveforms can be sent and thus multiplexing may be achieved. However, as actual values of signal amplitude are being sent the method is analogue and not digital transmission. Digital transmission derives from converting the discrete amplitude values into binary form, so that transmission is effected by sending a series of ones and zeros. The amplitude range is commonly represented by 256 discrete values and thus each sample uses 8 binary digits. If, as is generally the case, sampling takes place 8000 times per second, the transmission rate is 64 kbits/s.

Digital networks cannot be viewed in isolation as they will be required to interwork with analogue systems for many years, even though the final objective is an all digital network. The need to interwork with other national networks is a further constraint, as such networks may be more, or may be less, advanced than our own. There must clearly be uniformity of standards at international boundaries and this leads to a pressure, but not the necessity, for uniformity of standards within national boundaries as well.

Consequently, the voice requirements to be discussed are in accord with various CCITT recommendations detailed in the references. However, it is not surprising that, in North America, rather different standards are used in a network large enough for decisions to be made mainly in accord with internal requirements. Thus, it is sometimes necessary to make adjustments where the North American network has its interface with the rest of the world.

2.2 BASIC REQUIREMENTS FOR SPEECH - LOUDNESS LOSS

Speech heard over modern commercial telephone connections is, at its most basic quality, acceptably intelligible. Accordingly, the most basic need is for speech to be loud enough, but not too loud, for people with normal hearing acuity. It is in respect of achieving optimum loudness that digital transmission of speech introduces one of its chief advantages. With analogue networks, users experience speech which is sometimes too loud, and often too quiet, and thus the network planner has to ensure that the wide range of loudness levels occurring remain within a tolerable range.

In principle it is possible for a fully digital network to provide the same loudness level over all connections, both long and short. Before suggesting what this should be, it is necessary to discuss the means whereby it is determined. (2)

Naturally the loudnesss of speech heard by a listener depends upon the speech volume of the person with whom the conversation is taking place. This difficulty is avoided by using the concept of LOUDNESS LOSS. Loudness loss is independent of user behaviour, and is concerned with the ratio between sound pressure at defined points adjacent to the microphone at one end of a connection, and the earphone at the other. The mouth reference point (mrp) is 25mm in front of the plane of a talker's lips, and the ear reference point (erp) is at the centre of a plane formed by the rim of the earcap. Traditional earcaps were round, and the reference point easily defined, but this is no longer the case: more recent ideas tend toward defining a point at the ear canal entrance. In addition, it is necessary to define the talking position for handsets in relation to mrp, as this varies with the length and design of handsets. In practice, users hold handsets comfortably to their ear, and the position of the microphone, in relation to the lips, is dependent upon handset design. (1)

Telephone conversations may be considered as providing a substitute for face-to-face conversations to provide a starting point for discussing loudness loss measurement. The greater the distance between a talker's lips and the listener's ear, the greater the loudness loss will be. Of course two ears are used in a face-to-face conversation, but this difficulty is overcome by considering sound pressure at one ear only, both in the face-to-face case, and where a telephone link is used.

People engaged in a face-to-face conversation are typically one metre apart and, accordingly, the use of a one-metre air path reference system appears to have a conceptually realistic basis. A high quality electro-acoustical system, representing a one metre air path, was developed many years ago and used in two different ways to assess practical telephone connections. Both methods are now discredited, but a brief mention of them is necessary as their shortcomings are by no means obvious and the unwary may find them attractive. The earlier method was known as articulation testing, where sets of nonsense words were spoken at one end of a connection, and written down by a listener at the other. The proportion of sounds correctly recorded over the reference system and the practical system, provided a means of rating the system under test. The problem arises that a high proportion of sounds are correctly recorded for a circuit which may be described as poor, and thus hardly any higher score is achievable with good or excellent circuits. Hence, the method is quite insensitive for assessment of circuits of commercial quality. The method is used in areas where communication is difficult, such as under-water or in some military applications, but is of no interest now in telephony.

The other method not now used, involved comparing the loudness of speech heard over the reference system with that heard over a test connection. In this form the reference system is known as NOSFER. Extra loss was put into NOSFER so that equality of loudness was achieved between the test and reference connections. The amount of extra loss, in decibels, that was inserted was used as a measure of the performance of the test connection and known as REFERENCE EQUIVALENT. In practice reference equivalent values were obtained for parts of circuits (for reasons which will become clear below) and added together to give an overall value for a connection. This process introduces errors as large as 6dB and these arise from the fallacy of comparing narrow band-width (3.1kHz) practical circuits with a wide-band reference system.

The world telecomunications community, having used reference equivalents for many years, were reluctant to abandon them as units for network planning, even though their short-comings had been known for many years. As a short-term intermediate step, CORRECTED REFERENCE EQUIVALENTS (CRE) were introduced in 1981, to provide continuity with the vast amount of past planning decisions taken in reference equiv-alent terms. (5) CRE are reference equivalents, familiar after many years of use, weighted and modified to, at least in part, overcome the additivity problem. Following upon the last CCITT Plenary Session, the recommended units are LOUDNESS RATINGS which, in practice, have been used in the U.K. for some years past. It is of interest to note, how-ever, that the fundamental reference system remains NOSFER and that the reference system upon which loudness ratings are based is known as the INTERMEDIATE REFERENCE SYSTEM.

2.3   SENDING AND RECEIVING PERFORMANCE

With a public switched telephone network, it is insufficient to consider overall loudness loss, as there is a great variety of local systems which are interconnected. In practice a connection is considered as comprising three parts, a local system at each end and a junction in the middle. A local system includes the instrument, local line and the feeding bridge at the local exchange. The junction, which interconnects the two local ends, will vary widely bet-ween an 'own-exchange' call at one extreme and, at the other, may comprise plant necessary to provide a long international connection. The term 'junction' is used here in a general and not in a specific sense. The sending and receiving local systems will change roles frequently, perhaps several times per minute, during actual conversations.

Assessment is therefore made of the sending and receiv-ing performance of the local system up to, and including, the feeding bridge at the local exchange. The junction is treat-ed separately. The importance, and indeed the necessity, of this is made clear by considering interconnection between two national networks. If it so happened that one network prov-ided the desired overall loudness loss by using very sen-sitive microphones and correspondingly less sensitive ear-phones, while the other used less sensitive microphones and more sensitive earphones, then interconnection problems arise. By knowledge of the respective sending and receiving performances of different local systems a network may be planned appropriately.

Using the local exchange feeding bridge as a planning interface is in accord with U.K. practice. Elsewhere, e.g. in Sweden, the coresponding interface may be at the customer's premises and the local line treated as part of the 'junction'.

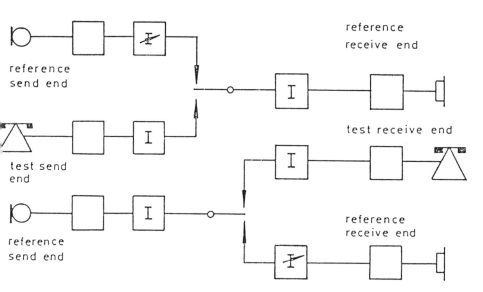

Fig.2.1 Measurement of sending and receiving
loudness loss

2.4 THE INTERMEDIATE REFERENCE SYSTEM (4)

If studies in this area were to commence today, it is very likely that the reference system which would be established would be very like what is now known as the Intermediate Reference System. It is usually called the IRS. It is called 'intermediate' because it represents a bridge between the Fundamental Reference System (NOSFER), based upon which there is a vast amount of historic planning data, and the practical connections it is necessary to consider at the present time. It is similar to NOSFER in that it is based upon high quality components and is tightly specified. However, it in no way represents an air path and is based upon a practical connection typical of those used in commercial telephony.

The main reason for the non-additivity of the old reference equivalent values was that they were obtained by comparing a wideband reference system with a practical connection employing 3.1 kHz/s bandwidth. This comparison between unlike circuits resulted in large errors. Because the intermediate reference system relates to standard practice, the results obtained by assessing parts of circuits may be added together to give an overall rating of a connection which is acceptably accurate. The units now used are LOUDNESS RATINGS.

Figure 2.1 is a quite general schematic diagram showing how, in principle, loudness ratings may be obtained. The upper diagram shows that the receive end of the reference system is switched between the send ends of the reference system and the send end of the connection being tested. A talker speaks at a controlled level into the two send ends while a listener to the receive end judges which is the louder. Adjustment is made to the attenuation of the reference send end until the listener judges the two paths to be equally loud. The amount of additional attenuation inserted into the reference send end to achieve this equality is known as the SEND LOUDNESS RATING (SLR). The lower diagram shows how, in a precisely similar fashion, the RECEIVE LOUDNESS RATING may be obtained (RLR). If the two values are added together a good approximation to the OVERALL LOUDNESS RATING (OLR) is obtained.

Because this statement is not absolutely true, and small errors in additivity remain, the term NOMINAL OVERALL LOUDNESS RATING (NOLR)is often used. The method described is equally applicable, in general outline, to the method used in the past to obtained the corresponding reference equivalent values. Loudness ratings have a further major advantage, apart from the matter of additivity. That is that they may be obtained by a calculation process which will be discussed later.

2.5   ADDITIVITY OF LOUDNESS RATINGS

The overall loudness rating of a connection usually comprises three parts: the sending and receiving ratings of the local systems and the loudness loss of a junction connecting them. Loudness loss of junctions is not the same as the attenuation which might be measured using pure tones. All the component frequencies of speech contribute to loudness and, therefore, the amplitude/frequency distortion of the plant comprising the junction affects the loudness loss. Additivity is straightforward only when this point is well understood. In practice pure tones are used to obtain an estimate of the loudness loss by measurement, but the choice of frequencies used will depend upon circumstances. It is clear that a single measurement carried out at say 800Hz is not adequate. When the network is fully digital there will be little difficulty, but a partly digital network, interworking with FDM systems, and both loaded and unloaded audio cable, is a much more complex situation.

It is fortunate, however, that additivity appears better if mismatches are ignored. It has been shown (13) that if other features of a connection element, such as amplitude/frequency distortion, are assumed to arise between (say) 600 ohm interfaces, the overall loudness loss obtained by simple addition, is very nearly the true value. This may be explained by the several mismatch losses which may arise having opposite signs and thus cancelling out.

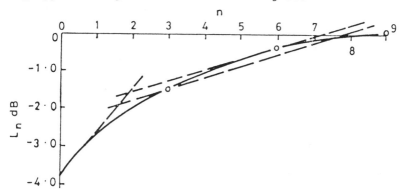

Fig.2.2 Loudness Loss L_n vs number of filter pairs n

The Intermediate Reference System uses two SRAEN filters in the junction connecting the two ends. A SRAEN filter represents three pairs of typical PCM or FDM filters. Figure 2.2 indicates how the loudness loss varies with the the addition of filter pairs in the junction. A straight line approximation is shown which is tangential to the curve where n = 6 chosen to be in accord with the IRS.

2.6 THE SIDETONE PATH

The loudness loss of another path must be considered, and that is the path, within one telephone instrument, from microphone to earphone. This is known as the sidetone path and arises because telephones are four-wire devices (two each for microphone and earphone) yet presenting two wires to line. A hybrid circuit connects the four-wire to the two-wire and, if this is not properly balanced, a sidetone path exists.

Assessment of the loudness loss of this path is exactly the same as has been discussed, except that it is considered overall, not in parts, taking due account that the two transducers are adjacent, rather than remote from each other. The Sidetone Loudness Rating (STLR) is obtained by comparing the overall loudness loss of the sidetone path with the IRS.

The matter is not, however, quite that simple. A telephone user hears his own voice via this path, but the electro-acoustical one, microphone to earphone, is not the

only path by which users hear themselves.   In addition there
is the human sidetone path which itself comprises two paths,
the acoustical one, and the path provided by the bone structure of the head.   To take account of this complexity,
Sidetone Masking Rating (STMR) has been introduced. (7)

Sidetone is conventionally regarded as a nuisance but
telephone users are accustomed to it, and feel something is
wrong with the telephone they are using if it is absent. A
four-wire telephone, such as might be envisaged on a fully
digital network, will have no inherent sidetone path and it
may be necessary to introduce one deliberately. The effect
of users hearing their own speech is to cause them to lower
their speaking volume, but this effect diminishes as sidetone
path attenuation is increased and has no effect when STMR is
greater than about 13dB. (10).

2.7 ASSESSMENT BY CONVERSATION METHODS.

Planning of telephone networks is, generally, on a
loudness loss basis. It is widely recognised, however, that
such methods do not assess the fitness of connections for
their primary purpose of enabling conversations to take
place. Modern thinking favours a conversation method of
assessment which includes many factors totally ignored by
loudness loss criteria.

Conversation assessment techniques were introduced long
before digital technology became a significant feature in the
network, but their use is particularly valuable in a digital
environment. Digital transmission changes the balance of
importance of traditional impairments. Loss, which in the
analogue environment is the most important single factor,
will cease to be a problem when digital transmission provides connections with optimum loudness loss performance
irrespective of length. A new impairment is quantizing
distortion. Conversation testing enables studies to be made
of the effects of the several impairments present, and thus
to re-assess their relative importance in a digital environment.

The units used in conversation testing are, in the main,
OPINION SCORES. The participants in a conversation test
are ordinary telephone users who are invited to converse over
a connection being assessed. The topic for conversation is
chosen so that neither participant is made dominant by the
nature of the conversation. This was a shortcoming of
early conversation test work, when puzzle solving tasks were
given to the participants. Nowadays it is common practice
to give each participant an identical set of pictures which
could be hung in a room and ask them to agree an order of
preference.

At the conclusion of each test, the participants are
invited to choose which of a predetermined set of descriptions best accords with their experience. They are asked

individually and not as a pair. Descriptions are of the form, excellent, good, fair, poor, bad, although other sets may be used (1). In addition subjects are asked whether they, or their correspondent, experienced any difficulty, and are restricted to a simple "yes" or "no" answer. As Practical network planning is aimed at minimising the number of customers who experience difficulty, this question is important, and is quite different from the description chosen. A subject may declare difficulty with a circuit described as good, and equally, no difficulty with a poor circuit. A note is made of the percentage of participants who experience difficulty and this is given the symbol %D.

The five-point opinion scale is numbered 4, 3, 2, 1, 0 from 'excellent' through to 'bad'. In the USA the scale is 5, 4, 3, 2, 1. Conversation tests involve a number of subjects, depending upon the size of the test but 24 is typical, and these will differ somewhat in their opinions. The opinion score is, therefore, the numerical average of the individual opinions.

In a public switched telephone network a very great variety of connections can occur, and it is not feasible to assess every one of them. Accordingly it is necessary to choose a number of connections which may be set up in the laboratory and are typical of the population of connections which may occur. Such connections are known as HYPOTHETICAL REFERENCE CONNECTIONS (HRC). The selection of HRC, suitable for a given conversation test, depends upon the features being studied. With the wide ranges of transmission loss, line noise, room noise, sidetone ratings etc. which can occur, many possibilities may be envisaged. It is necessary to fix the values of most quantities in order to study the effect of varying others. Otherwise, tests would take far too long to carry out but unfortunately interactions between features to be studied, with those that are fixed, become obscured.

Subjective tests of all kinds are very expensive, and laborious, this being particularly true of conversation tests. Thus, only the largest laboratories are able to devote adequate resources to the matter. Even more serious, however, is the long delay between a test being envisaged, and the results being available. This mitigates against the use of subjective tests for optimising between several interacting factors. Accordingly, calculation methods are now widely used, both to obtain loudness ratings and opinion scores. It must, however, be remembered that the information used in calculation procedures has, at some earler time, been obtained from tests using subjects.

2.8 CALCULATION METHODS

Calculation of loudness ratings is relatively straight-forward (1,6). A description of the Intermediate Reference

System (IRS),and a detailed specification of the test connection, are required.   The test circuit specification includes sending and   receiving sensitivities   of the   telephone sets, obtained in a   defined way. CCITT Recommendation P64 (8) describes   the method to be used in obtaining set sensitivities. Also,   a description of the   electrical elements is required. The approach used in USA is described in IEEE Standard 269.

The total bandwidth is   divided   into   narrow   bands   of frequencies, and the contribution   of   each   to   loudness   is suitably   weighted, before summation to   give a loudness numeric.   This   is done, in   principle,   for both   the IRS, and the practical   test connection,   and the   difference between the two   loudness numerics enables the   loudness rating to be determined.   In practice the IRS is defined, and the loudness numeric   for it already   known, and built   into the procedure for obtaining the loudness rating for the test circuit.

In principle, calculation of   loudness ratings is quite simple   because it   depends upon   sets of   objective electroacoustical measurements.   Calculation of opinion scores is a more complex   matter, as   it involves   assumptions concerning human telephonic behaviour, and aims to predict the responses of actual subjects, talking over a real circuit.   One of the problems arising   in calculation procedures   is the looseness of coupling of the earphone to a user's ear.

Earphone sensitivities are   usually   obtained   with   the earphone sealed to an artificial ear, and this is allowed for in   the calculation procedure by use of a frequency dependent correction to   take account   of loose   coupling at   the human ear.   These   are known as   'real ear losses'   , published by CCITT, and reproduced in Table 1.   It is desirable that such figures   be   an   accurate   representation   of   actual   user behaviour, but it   is quite   essential that   the same   set be used by each Administration to ensure consistency in loudness ratings quoted by them.   Accordingly,   other sets of real ear loss figures which   have resulted   from more   recent studies, are agreed to be more accurate, and suitable for research and calculation of   opinion   scores, but   not   applicable   to calculation of loudness ratings until universally adopted.

Calculation   of loudness ratings   requires definition of the   interfaces   between   which   the   ratings   apply.   In particular, it is necessary to define digital interfaces, and digital sequences   at   these,   corresponding   to   appropriate analogue   values.   An obvious   point is that   signals are in analogue form at the interface with the user.   Therefore, if transmission   over the local system is digital, the telephone set will   incorporate a   codec and   thus there   is a   need to relate   digital sequences to sound   pressures.   Transducer sensitivities (microphone and earphone alike) are given as a relationship, at   each   frequency, between voltage   at   the terminals,   and sound pressure at   defined points adjacent to the diaphragm. CCITT Recommendations G711 and G101 define the digital sequence which, when fed into a   specified decoder, provides an analogue dBmO. If 600 ohm impedances are assumed, the value in dBVO is taken to be 2.2dB higher.

If a sound source is applied in accord with CCITT recommendation P64 (8) the Send Loudness Rating is given by the expression

$$SLR = - 10/m \ \text{Log}_{10} \ 10^{m/10 \ (S_{MJ} - W_S)} \ dB$$

where m is 0.175 and depends upon the slope of the loudness growth function (1).

S_{MJ} is, at each frequency, the sensitivity between the mouth reference point and the junction discussed earlier.

W_S is the weight attached to each sensitivity value S_{MJ}, to provide the contribution to total loudness of each one-third octave band represented by the chosen frequencies.

Similarly the expression for Receive Loudness rating is:

$$RLR = - 10/m \ \text{Log}_{10} \ 10^{m/10(\ S_{Je} - L_E - W_R)} \ dB$$

where S_{Je} is the sensitivity between the junction and the ear reference point assuming the earphone sensitivity has been obtained when sealed to an artificial ear. If the sensitivity were obtained with a real ear, the symbol S_{JE} would be used and the real ear loss term L_E not be included.

L_E is the value for real ear loss referred to earlier.

W_R is the weighting for receiving corresponding to W_S.

The Sidetone Masking rating incorporates the effects of the several paths discussed earlier. Acoustic loss measurements are made between mouth and ear reference points and the effects of other paths included in the weights W_M. As the loudness growth function is different in this case the value for m is changed.

$$STMR = - 10/m \ \text{Log}_{10} \ 10^{m/10 \ (-L_{MeST} - L_E - W_M)} \ dB$$

where m = 0.225

L_{MeST} is the loss, in dB, between mouth and ear reference points assuming an artificial ear has been used in determining earphone sensitivity.

W_M is the weighting for each band corresponding to W_S and W_R.

TABLE 2.1

WEIGHTS USED IN CALCULATING LOUDNESS RATINGS

Frequency (Hz)	W_S (dB)	W_R (dB)	W_M (dB)	L_E (dB)
100	154.5	152.8	94.4	20.0
125	115.4	116.2	90.4	16.5
160	89.0	91.3	89.8	12.5
200	77.2	85.3	86.4	8.4
250	62.9	75.0	81.9	4.9
315	62.3	79.3	78.5	1.0
400	45.0	64.0	78.3	-0.7
500	53.4	73.8	72.8	-2.2
630	48.8	69.4	67.6	-2.6
800	47.9	68.3	58.4	-3.2
1000	50.4	69.0	49.7	-2.3
1250	59.4	75.4	48.0	-1.2
1600	57.0	70.7	48.7	-0.1
2000	72.5	81.7	50.6	3.6
2500	72.9	76.8	49.8	7.4
3150	89.5	93.6	48.4	6.7
4000	117.3	114.1	49.2	8.8
5000	157.3	144.6	48.3	10.0
6300	164.7	152.2	48.1	12.5
8000	151.7	136.7	51.7	15.0

2.9   COMPUTER MODELLING

Rapid calculation of loudness ratings may be carried out by use of a suitable computer program. Some information such as the weights listed in Table 1 may be built into the program. Other information, such as the sensitivities of commonly used telephone sets, may be stored in files associated with the program and called upon as required, while yet other data, unique to the calculation being made, fed in while the calculation is in progress.

The best known of the computer models is CATPASS (computer assisted telephone performance assessment), now used in the more useful forms of CATNAP (computer aided network planning programm) and TCAM (telephone connection assessment model). CATPASS and CATNAP are due to British Telecom, TCAM is a development carried out at the University of Aston. The ATT transmission rating model is conceptually different and not considered here. The models allow for calculation of loudness ratings, opinion scores, and the percentage of persons who will experience difficulty.

Modelling requires the use of information derived from experiments, such as that represented by Table 1. The models comprise several stages, and TCAM is constructed to make this explicit. Early stages deal with objective descriptions of the connection elements which are cascaded. The next stage is calculation of loudness ratings, and the next with determining the noise level at the ear reference point. Final stages involve prediction of opinion scores, based upon earlier conversation tests.

Models used to predict the opinion scores to be obtained in a conversation test must be dynamic, and be updated to allow for changing, usually rising, expectations of telephone users. Because this is so, models tend to predict scores somewhat higher than actual tests provide. A distinction is necessary between models such as CATNAP, used for network planning, where consistency requires the model to be unchanged for a number of years, and research based models, such as TCAM which is frequently updated as new knowledge is obtained.

The real test of any connection is customer satisfaction after installation of the plant. Mistakes, leading to substantial customer dissatisfaction, are very expensive to correct and laboratory testing prior to installation is essential. Planners of the analogue network have many years of accumulated experience behind them, but the digital network presents new problems and requires new knowledge. It is in this respect that computer models are particularly valuable. To carry out subjective tests covering the whole range of features that may arise, and studying their interactions, is a daunting prospect. However, computer models allow 'notional tests' to be undertaken very quickly and cheaply, enabling suitable selections to be made from the wide range of possible hypothetical reference connections for use in actual subjective tests. The models, therefore, provide a guide to what is important and what is less so and, accordingly, while decisions are based upon evidence from actual tests, the choice of what to test is greatly assisted by the models. However, modelling is an iterative process for, if actual tests relate well to model predictions, confidence in the model is enhanced and it may be relied upon to predict results of small variations from conditions actually tested.

2.10 REDUCED BIT RATE SYSTEMS (14, 15)

Conventional 64 kbit/s systems provide a very high quality transmission path for speech. However, they are not really efficient as it is possible to provide adequate transmission quality with very much less capacity. This is because of redundancy in the speech arising from the slow rate of change of the envelope.

Many administrations have been investigating this matter in recent years and a number of different approaches adopted. Mobile radio telephony provides a current practical example of the use of bit reduced systems in the field, the use of which which is becoming increasingly widespread. In addition, 32 kbits/s ADPCM (adaptive differential pulse code modulation) systems have been under study for some years. The voice requirements of all these systems are, of course, precisely those required of older technologies in the past, namely that they should provide commercial quality connections for the user to converse over. Providing this basic requirment is met, it is a proper aim to increase the capacity of systems as far as is economic and practical and, no doubt, this trend will continue.

The matter of assessment and evaluation of these newer systems is of great importance. The aim of work in this area is to develop an objective measure of circuit quality which can be related to the subjective experience of users. This is not an easy task and the objective has yet to be achieved.

One approach is to let the quantizing distortion produced by a conventional 64 kbits/s PCM system be known as one quantizing distortion unit (QDU). It may be then that a bit reduced system will produce 2 or 3 QDU. On the assumption that QDU are additive, the total number of QDU produced by a number of systems in tandem is obtained by adding together the QDU produced by each component system. If the total does not exceed the allowable maximum then the overall connection is satisfactory. It follows that the more QDU a particular type of system produces, the smaller is the number of such systems that can be put in tandem.

This approach is the only one possible at the present time but it is not really valid. What is required is a generally applicable objective measure which relates to subjective experience over a wide range of conditions. The only generally available measure for assessment of bit reduced systems consists of listening tests and these are in use in laboratories.

The problem is that if a relationship between objective and subjective measures is achieved for a system this does not hold when a number of such systems are interconnected. Further, if the relationship be established for a system it is found to hold only under the conditions used for its determination. Any changes, such as a variation in talker level, invalidates the relationship. Accordingly, the use of QDU and merely adding them together, is a practical technique to provide guidance in network planning. It is not strictly valid.

Assessment of 64 kbits/s PCM systems has for a long time involved use of a device known as the modulated noise reference unit (MNRU). In this device, both random noise and speech are fed to a modulator which is so configured that the noise level is proportional to speech amplitude. Various settings of the relationship between noise and speech are possible, and these hold over a wide range of speech levels. The device enables objective measurements to be made of the performance of PCM systems. (1)

2.11 DIGITAL TELEPHONE SET CHARACTERISTICS

Performance requirements of speech links have evolved over many years and it is unlikely that the introduction of digital technology will suddenly change them. Digital technology will make their achievement much easier and raise the quality of the service the public enjoys to a level always desired, but only exceptionally achieved. Therefore, the frequency response of telephone sets in a digital environment will be in accord with the best practice on the analogue network. (2) A suggested characteristic given in Table 2 corresponds to the long term objectives contained in CCITT Recommendation G.111.

The analogue and digital parts of connections may be defined separately in order that their combined performance is satisfactory, but this may be uneconomic. There is no purpose in tightly specifying an element in the analogue part, if the digital part is totally dominant in a particular regard. For example, the sharp upper-frequency cut-off introduced by codec filters makes audio frequency cut-off immaterial. On the other hand, if the sidetone path is totally analogue, audio frequency cut-off cannot be ignored.

2.12 CONCLUDING COMMENTS

The implicit assumption that echo effects are negligible has been made throughout this discussion. Clearly, such an assumption is not always valid. In addition, treatment of sidetone and noise has been cursory. Related to sidetone is the matter of stability, for the acoustic coupling between microphone and earphone can cause 'howling', particularly if a handset is placed upon a hard surface. Accordingly, singing margins must be specified.

Users hold their handset as closely to the ear as they feel necessary, and this seems to depend upon the perceived speech to noise ratio. If the speech is loud, the earphone is held loosely, which reduces the room noise heard via the sidetone path, but increases the amount entering the leakage path at the pinna. When the speech to noise ratio is adequate, the user has little incentive to hold the earphone more closely to the ear even if this improves the ratio. Adaption by users to various conditions is a factor to be taken into account, and it may be concluded that excessive loudness is pointless as users merely discard it by movement of the earphone away from the pinna. (11, 12).

Although a completely digital network will make aiming for a single figure (rather than a range of values) for overall loudness loss a feasible possibility, this is neither necessary nor desirable. An increase of (say) 4dB in the preferred loss will have negligible effect upon users' opinions of a connection, when it is echo free, and may be preferred if echo is present.

TABLE 2.2

SENSITIVITIES FOR A DIGITAL TELEPHONE SET
(referred to an interface of 0dB relative level)

Frequency (Hz)	Send Sensitivity S_{MJ} dBV/Pa	Rec. Sensitivity S_{Je} dBPa/V	Sidetone Loss L_{MeST} dB
100	-44.8	-23.4	66.7
125	-35.8	-15.5	49.7
160	-26.8	- 8.8	34.1
200	-21.0	- 1.4	20.9
250	-17.5	2.6	13.4
315	-15.0	6.2	7.3
400	-13.3	8.1	3.8
500	-12.7	8.7	2.5
630	-12.3	8.9	1.9
800	-11.3	9.0	0.9
1000	-10.1	9.3	-0.7
1250	- 8.9	9.1	-1.7
1600	- 7.6	9.1	-3.0
2000	- 7.3	8.7	-2.9
2500	- 6.3	8.2	-3.4
3150	- 7.5	8.5	-0.5
4000	-33.2	-25.4	58.1
5000	-78.3	-77.9	155.0
6300	-86.1	-92.0	178.0
8000	-94.5	-100.3	200.0
Loudness Ratings:	SLR = 6.0	RLR = 3.0	STMR = 9.3

ACKNOWLEDGEMENTS

    I am greatly indebted to Dr. D. L. Richards for
discussions about matters contained in this brief survey and
for his guidance over many years, and also to Mr. Brian
Surtees of B T Research Laboratories for useful discussions.
I am totally responsible for any omissions or errors which
may occur.   To those who need to know more, I commend Dr.
Richards' book 'Telecommunication by Speech' which covers
many aspects in great detail, and introduces the reader to
further information by its comprehensive bibliography.

REFERENCES

1.   Richard, D. L., 'Telecommunication by Speech',
      Butterworth 1973.

2.   Richards, D. L. and Whorwood, R. W., 'Transmission
      Planning of Digital Telephone Networks', ISSLS-80
      Munich, Germany.

3. CCITT Recommendation P.72 'Measurement of Reference Equivalents and Relative Equivalents', Orange Book V, 1977.

4. CCITT Recommendation P.48 'Specification of an Intermediate Reference System', Orange Book V, 1977.

5. CCITT Recommendation G.111, Annex 1, 'Definition and Properties of the CRE's', Yellow Book III, 1981.

6. CCITT Recommendation P.79 'Calculation of Loudness Ratings', Yellow Book V, 1981.

7. Hoppitt, C. E., 'Telephone Sidetone Ratings which includes Human Sidetone Effects', Electronic Letters, 18 March 1976.

8. CCITT Recommendation P.64 'Determination of Sensitivity/Frequency Characteristics of local telephone systems to permit calculation of their loudness ratings', Yellow Book V, 1981.

9. Richards, D. L. and Webb, P. K., 'CATPASS - model for estimating customer satisfaction', Conference Record, NTC - 1976.

10. Coleman, A. E., 'Sidetone and its effect on Customer Satisfaction', British Telecommunications Engineering April 1982.

11. CCITT Study Group XII, Contribution COM XII - No. 95 (Hungary) Period 1977-1980.

12. Richards, D. L. and Whorwood, R. W., 'Acoustical Transmission Losses in Telephone Connections' Proc. Inst. Acoustics, 1981.

13. Richards, D. L., Private Communication, 1982.

14. CCITT Recommendation G 721. Red Book, 1985.

15. Supplement 14 to 'P' Series Recommendations, Vol V Red Book 1985.

Chapter 3

Data transmission

Dr Ron Brewster

3.1   INTRODUCTION

Data transmission is the process of communicating information in digital form from one location to another. If the communication is simply between two terminals in reasonably close geographical proximity, then there is really no problem in providing a dedicated connection with adequate bandwidth and suitably chosen characteristics to enable direct interconnection to be made. Difficulties arise, however, where there are several terminals that need to be interconnected so that communication can take place between any pair (or more) of terminals, or where the distances involved, and the usage is such, that it is uneconomic to provide dedicated special-purpose circuits. The first of these problems may be sub-divided into two categories. Firstly there are those data networks which are geographically confined to a fairly small area, such as a factory site, university campus or office block. Such networks are known as Local Area Networks (LANS). Secondly there are also in existence data networks which cover whole geographical regions, from industrial conurbation to national, or even international, operations. Clearly the Wide Area Networks (WANS) embrace not only the problems of terminal interconnection protocols but also those of long-distance data transmission. We shall address ourselves in this chapter to the problems of long distance transmission.

3.2   DATA TRANSMISSION OVER PUBLIC TELEPHONE NETWORK

The provision of telecommunications transmission facilities beyond the boundaries of the Local Area Network has, until recently in the UK, been the sole prerogative of British Telecom (formerly BPO). Even with the emergence of new carriers, such as Mercury, the very ubiquitousness of the telephone network makes it an attractive candidate for long-distance data transmission. It was, of course, originally designed to carry analogue speech signals and lacks many of the properties desirable in a data network. Techniques have been developed, however, to overcome most of the shortcomings and data transmission facilities are now available both nationally and internationally over the public telephone network. In the UK these services are designated the DATEL services.

The major shortcoming of the telephone network for data transmission is that the network is designed to transmit speech signals which occupy a bandwidth from about 300 to 3400 Hz. Since speech is not significantly impaired by phase distortion, no attention is paid to the network phase-versus-frequency characteristics. The d.c. signal path is removed by exchange bridges and a.c. coupling, although it exists on local exchange lines for network signalling purposes. The cut-off at higher frequencies occurs due to loading coils, frequency-division-multiplex filters and p.c.m. anti-aliassing filters. Since the spectrum of a data signal may, in principle, possess significant d.c. and low-frequency components, it is necessary to shift the spectrum of the base-band signal using modulation. The modulator and demodulator necessary for this process are provided as line-terminating 'black box' referred to as a Modem (MOdulator - DEModulator).

The first modems to be introduced into service transmitted binary data by means of frequency modulation. Since switching between two frequency states only is required, the technique is generally referred to as frequency-shift-keying (FSK). FSK was chosen because it gave a good performance, was reasonably simple to implement, provided stable operation over a wide variety of channels and permitted asynchronous operation. Additionally, the modem performance is not affected in any way by the data sequence because the line signal is always present with constant amplitude whatever the pattern of binary 1s and 0s. Two standard FSK systems are in general use. The first was designed to enable keyboards and similar machines to communicate over the public switched telephone network (PSTN) and private speech circuits at signalling rates of up to 200 bit/s in a full duplex mode. CCITT recommendation V21 was established to cover this requirement internationally and made provision for a possible extension of the rate up to 300 bit/s in each direction. The frequency allocation chosen is given in table 1.3. Asynchronous or synchronous working between data terminal equipments is possible when using these modems. The UK version of this modem is known as Datel Modem No.2 and in the USA the Bell System Dataset No.103 provides similar facilities.

TABLE 1.3   Characteristic frequencies for V21 modems operating at 200 bit/s

Channel	Nominal Mean Frequency F_o (Hz)	Binary Symbol 1 Frequency F_z (Hz)	Binary Symbol 0 Frequency F_a (Hz)
1	1080	980	1180
2	1750	1650	1850

Channel 1 is used for the transmission of the caller's data, while channel 2 is used for transmission in the other direction.

The second standard was introduced to provide a higher speed link, the main use of which has been to enable visual display units to be serviced by a host computer with greater rapidity than was possible by the previously described modem. A low-speed return channel, which would allow control signals or telegraph-type keyboard signals to be transmitted was provided as an option if required. The modem again operates on either the PSTN or private speech circuits, this time at data signalling rates of up to 1200 bit/s. However, with a small proportion of the poorer quality switched network connections, difficulty was experienced in obtaining 1200 bit/s and so a fall-back facility was provided to enable operation at 600 bit/s to be assured over the PSTN. The return channel operates at modulation rates of up to 75 bit/s. The frequency allocation for this modem is given in table 2.3. The international standard is established in CCITT recommendation V23. In the UK, Datel Modem no 20, which supercedes Modem No1 conforms to this recommendation and in the USA Bell System Data Set No. 202 fulfills the same requirement.

TABLE 2.3 Characteristic frequencies for V23 modems operating at 600/1200 bit/s

Mode	Nominal Mean Frequency F_o (Hz)	Binary Symbol 1 Frequency F_z (Hz)	Binary Symbol O Frequency F_a (Hz)
1 (Up to 600 bit/s)	1500	1300	1700
2 (Up to 1200 bit/s)	1700	1300	2100
Optional Return Channel (75 bit/s)	420	390	450

For data signalling rates in excess of 1200 bit/s more complex techniques have to be employed. There are a number of ways of increasing the data signalling rate for a given bandwidth and these may be used individually or in combination. Firstly, more efficient modulation systems which require less line bandwidth for a given modulation rate can be employed. Examples of these are single-side-band (SSB) amplitude modulation, vestigial-side-band (VSB) amplitude modulation and quadrature amplitude modulation (QAM). In this latter system the data is divided into two streams which are separately double-side-band suppressed carrier amplitude modulated onto two carriers at a single frequency but in a phase-quadrature relationship.

Secondly, multilevel modulation may be used whereby the degree of modulation of the line signal is determined from consideration of more than one bit of data. If n bits of data are considered at a time there are 2^n possible different combinations that can be received from the data terminal equipment. Each of these 2^n states may be made to modulate the line signal a different amount. Thus in an FSK system the transmitted signal may be shifted to any one of 2^n different tones, or in an AM system to any one of 2^n different levels, or in a phase-shift keyed system (PSK) to any one of 2^n different phases. It can be seen that by this method the line signal modulation rate (Baud rate) is only 1/n th of the data signalling rate. Thus the line bandwidth required is reduced by a factor n compared to that required for binary modulation. With all of these more complex modulation techniques it is generally necessary to equalise the line or channel characteristics and, especially for the higher data rates, this often has to take the form of automatic adaptive equalisation, which is able to adjust to any changes in the channel characteristics as they occur.

The first requirement for a data rate greater than 1200 bit/s was met by a modem designed to operate at 2400 bit/s. Originally this modem, to CCITT recommendation V26, was intended for operation over 4-wire private speech band circuits. However, a variant (V26 bis) was subsequently introduced for use over the PSTN. This variant was provided with a fall-back facility to 1200 bit/s where the line characteristics were unsuitable for 2400 bit/s operation. The modem operates at a modulation rate of 1200 bauds, the data being taken in pairs of bits and being conveyed as one of four possible phase-shifts of a 1800 Hz carrier. The 4-phase signal

is differentially encoded to avoid the need to transmit a reference phase signal for the purposes of demodulation. The information is thus decoded at the receiver by comparing the phase of the carrier preceding the instant of modulation with that following it. The significance of phase changes is given in table 3.3. The two alternatives are available for the leased line modem (V26) but the switched network modem (V26 bis) is only available using alternative B. The advantage of alternative B is that there is always a phase-change between adjacent modulation epochs.

TABLE 3.3 Line signal phase changes for differential 4-phase modem

Pair of Data Signal Elements (dibit)	Phase Change	
	Alternative A	Alternative B
00	0°	$+ 45^o$
01	$+ 90^o$	$+ 135^o$
11	$+ 180^o$	$+ 225^o$
10	$+ 270^o$	$+ 315^o$

The nature of the 4-phase differential modulation technique used in this modem in which pairs of data signal-elements are identified by phase changes of the carrier signal, requires that the transmission be synchronous, ie that both the modulator and demodulator, together with the Data Terminal Equipment, must be controlled by timing signals in such a way that the individual data-signal elements can be correctly identified in the demodulation process. Within the UK the Datel Modem No 12 and in the USA the Bell Data Set No201 conform with this recommendation.

Following the successful operation of 2400 bit/s modems, the next requirement is for 4800 bit/s operation. This is achieved by the use of differential 8-phase modulation on a 1800 Hz carrier at a modulation rate of 1600 bauds. Equalisation is necessary in order to obtain satisfactory operation at this data rate. Three alternatives have been recommended by CCITT as follows:

V27   4800 bit/s modem with manually set equaliser for use on high quality leased voice-band circuits.

V27 bis   4800 bit/s modem with automatic adaptive equaliser and fall-back facility to 2400 bit/s operation in accordance with recommendation V26A for use on general quality leased voice-band circuits.

V27 ter   4800 bit/s modem with automatic adaptive equaliser and fall-back facility to 2400 bit/s operation in accordance with recommendation V26A for operation over the PSTN.

The significance of the phase changes for each of the options is given in table 4.3.

TABLE 4.3   Line signal phase changes for differential eight-phase modems

Tribit Value	Phase Change
001	0^{0}
000	45^{0}
010	90^{0}
011	135^{0}
111	180^{0}
110	225^{0}
100	270^{0}
101	315^{0}

To increase the data rate further using PSK would involve the use of 16 phases. The noise margin for 16 phase operation is, however, extremely poor. The noise margin can be greatly improved by the use of combined phase and amplitude modulation. CCITT recommendation V29 describes a 9600 bit/s modem for use over leased voice-band circuits using this technique. The signal space diagram for this modem is given in Fig 3.1.

The modulation rate is 2400 bauds. Fall-back rates of 7200 and 4800 bit/s are provided. In each case the modulation rate is kept at 2400 bauds, the number of points in the signal constellation being reduced to 8 and 4 respectively. Again, automatic adaptive equalisation is necessary for satisfactory operation.

3.3   BASE-BAND MODEMS

Not all data transmission involves the use of the telephone network. In many applications, especially in the local area, private wires are provided to carry data. Such circuits are normally low-pass rather than band-pass and modulation is not therefore required to translate the signal into the pass-band. However, some signal processing is usually desirable and, although the processes of modulation and demodulation are not included, it has become the custom to refer to the line terminating equipment provided to carry out this processing as a base-band modem.

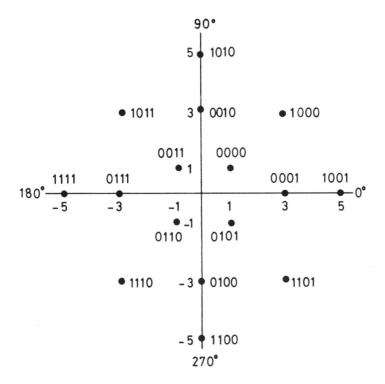

Fig 3.1 SIGNAL SPACE DIAGRAM FOR 9600 bit/s V29 MODEM

The base-band modem usually performs two functions. Firstly it converts the binary data input stream into a suitably encoded line signal to ensure that adequate timing information is contained in the transmitted signal and that the signal energy is sensibly distributed over the channel band-width. Secondly it provides channel filtering so as to reduce the interference between adjacent transmitted pulses (intersymbol interference).

3.3.1 Line codes

The simple binary bipolar code, where 1 is represented by a positive impulse and O is represented by a negative impulse, and the simple

unipolar binary code, where 1 is represented by a positive impulse and O is represented by no signal, both suffer from the existence of strong frequency components at low frequencies and a lack of signal transitions when long strings of Os and ls are transmitted. These transitions are necessary to derive timing signals for the receiver decoder. Their basic nature, however, make them a useful reference of comparison for more complex code arrangements. Two modified binary codes are the Walsh function codes Wal 1 and Wal 2 (Duc and Smith (1)). These are illustrated in Fig.3.2 (a). In both cases the line signal is free from d.c. component and contains a large number of transitions from which timing information

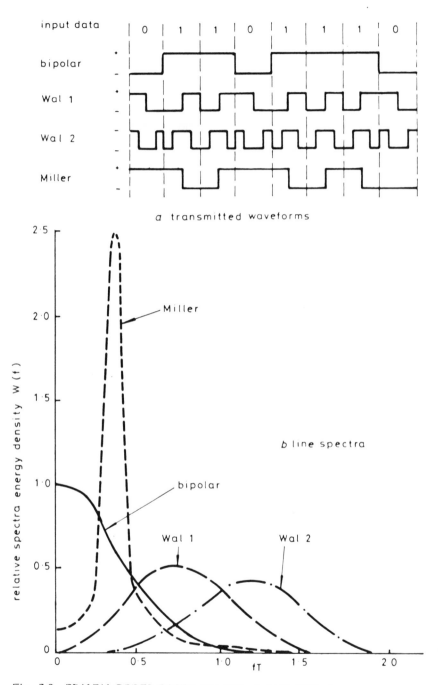

a transmitted waveforms

b line spectra

Fig. 3.2 BINARY CODES BASED ON WAL 1 AND WAL 2

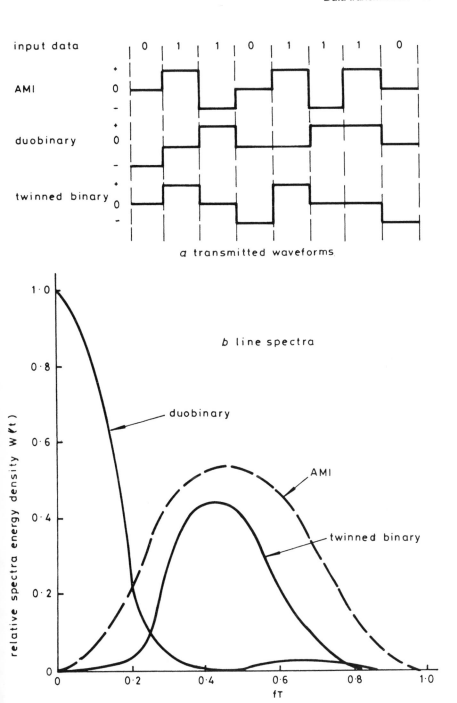

Fig. 3.3 AMI AND PSEUDO-TERNARY CODES

can be recovered, whatever the transmitted data pattern. The line spectra for these two codes are shown in Fig.3.2 (b). In both cases the spectrum has no d.c. component and low l.f. components. the band-width occupancy of the Wal 2 is slightly greater than the Wal 1, extending to approximately twice the transmission rate. The Wal 2 spectrum has no significant content below O.2 of the transmission rate, a fact that can be made use of in data-over-voice applications. Another useful binary code is the Miller code, or delay modulation, as it is sometimes called. This code is a variation of the Wal 1 and is derived by deleting every second transition in a Wal 1 signal, as illustrated in Fig.3.2 (a). The spectrum shape of the line signal is given in Fig.3.2 (b), where it can be compared with the Wal 1 and Wal 2 spectra. Although it has a small d.c. component, it has the advantage of a more limited band-width requirement than the comparable Wal codes.

Ternary codes operating on three signal levels, the middle level of which is usually zero volts, have found wide application as line codes. The best known of the ternary codes is alternate mark inversion (AMI). The encoded sequence is obtained by representing the mark in the binary sequence alternately by positive and negative impulses whilst the spaces are represented by no signal. The AMI code has very attractive properties. The line signal power density spectrum has no d.c. component and very small l.f. spectrum content. The coding and decoding circuitry requirements are quite simple and some degree of error monitoring can be achieved by simply observing violations of the AMI rule. The band-width required is equal to the transmission rate. It has the disadvantage that it has a poor timing content associated with long runs of binary zeros. There are two useful ternary codes which fall within the sub-class of linear pseudo-ternary codes (Crosier (2)). These codes are designated pseudo-ternary since they are determined by a set of rules which assign a three-level signal to a binary message and linear because the pseudo-ternary code is linearly derived directly from the binary message. A linear pseudo-ternary code is actually a particular case of a binary code in which the signal element $S_0(t)$ has been replaced by

$$S'_0(t) = S_1(t) \times S_0(t)$$

Where $S_1(t)$ is the sequence of impulses

$$S_1(t) = \sum_{k=0}^{K} \propto_k \delta(t - kT)$$

A linear pseudo-ternary encoding is thus equivalent to a filtering operation, the frequency response of the equivalent filter being given by the Fourier transform of S_1

$$S_1(w) = \sum_{k=0}^{K} \propto_k e^{-jwkT}$$

In order for the coded signal to have only three possible values for any sequence it is necessary that there be only two $\propto_k \neq O$ and that they are either equal or opposite. We thus have two basic pseudo-ternary codes; the twinned binary in which $\alpha_0 = -\frac{1}{2}$ and $\alpha_1 = +\frac{1}{2}$ (or the other way round) and the duobinary code, in which $\alpha_0 = \alpha_1 = +\frac{1}{2}$. The AMI and pseudo-ternary codes are illustrated in Fig.3.3, together with the power spectra for random data using these codes. The duobinary code has all its energy concentrated at low frequencies and has a very strong d.c. component. However, since the significant spectrum band-width is equal to only half the data rate, the code is attractive for use in limited bandwidth applications. The twinned binary code band-width is equal to the transmission rate and most of the signal energy is concentrated around half the bit rate. The spectrum is thus very similar to that of AMI code. The code is easily generated and has an error-detecting capability since it obeys the AMI rule. The code is, however, sensitive to error in the decoding operation due to error propagation in the decoding circuit. There are also two classes of non-linear ternary codes which have application in the field of data transmission, namely alphabetic and non-alphabetic codes. In the alphabetic codes n binary digits are taken together giving a signal element which can be regarded as a selection from an alphabet of 2^n possible characters. The character is then encoded into m ternary digits where $3^m > 2^n$. Such codes are normally described as nNmT codes. The simplest of these codes is given by n = m = 2 and is generally known as Pair Selected Ternary (PST). The message signal is grouped in 2-bit words which are then coded in ternary as given in table 5.3.

TABLE 5.3   PST Code Translation

| Binary Word | Ternary Word | | Word Digital Sum |
	Mode A	Mode B	
0 0	- +	- +	0
0 1	0 +	0 -	± 1
1 0	+ 0	- 0	± 1
1 1	+ -	+ -	0

It can be seen there is no change in the rate of transmission. With this code the d.c. component is reduced by the alternating mode which is equivalent to the alternating polarity of bipolar coding. Timing content is assured by the translation of pairs of zeros into pulses. The mode alternation also provides some error monitoring capability. The average power spectrum is given in Fig.3.4(a). The drawbacks of the code are that it has high l.f. components and the transmission power is about 1.5 times that of AMI for similar performance.

A widely used alphabetic code is that known as 4B3T. The original binary data stream is divided into words of four bits, each word being encoded into three ternary digits, as shown in table 6.3. The average power spectrum of the 4B3T code with random data is given in Fig.3.4 (a). The power is fairly evenly distributed throughout the spectral band but there is a significantly large component at the low frequency end of the

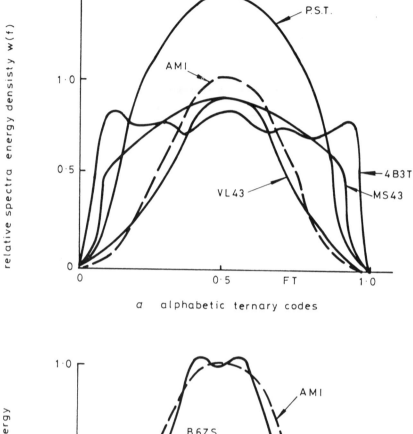

a alphabetic ternary codes

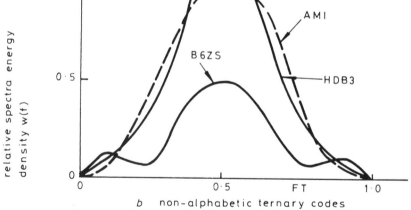

b non-alphabetic ternary codes

Fig. 3.4 POWER SPECTRA FOR TERNARY CODED RANDOM DATA

spectrum. Some attempts to overcome this large l.f. content have been made by the introduction of modified 4B3T type codes. Two of these are the MS-43 code and the VL-43 code. The average power spectra for these two codes are also given in Fig.3.4 (a).

In the non-alphabetic codes, long runs of zeros which may occur in conventional AMI coding are broken up by the substitution of pulses or groups of pulses which violate the AMI alternating mark pulse polarity rule. There are a number of ways in which this may be carried out and the most useful of these are described in (2). The average power spectra for the two most important non-alphabetic codes are given in Fig.3.4 (b). These are the B6ZS (Binary with Six Zeros Substitution) and the HDB3 (High Density Binary with maximum of three consecutive zeros).

TABLE 6.3 4B3T code translation

Binary Word	Ternary Word		Word Digital Sum
	Mode A	Mode B	
0 0 0 0	+ 0 -	+ 0 -	0
0 0 0 1	- + 0	- + 0	0
0 0 1 0	0 - +	0 - +	0
0 0 1 1	+ - 0	+ - 0	0
0 1 0 0	+ + 0	- - 0	±2
0 1 0 1	0 + +	0 - -	±2
0 1 1 0	+ 0 +	- 0 -	±2
0 1 1 1	+ + +	- - -	±3
1 0 0 0	+ + -	- - +	±1
1 0 0 1	- + +	+ - -	±1
1 0 1 0	+ - +	- + -	±1
1 0 1 1	+ 0 0	- 0 0	±1
1 1 0 0	0 + 0	0 - 0	±1
1 1 0 1	0 0 +	0 0 -	±1
1 1 1 0	0 + -	0 + -	0
1 1 1 1	- 0 +	- 0 +	0

Codes using more than three pulse amplitude levels are possible. For instance, quarternary (four-level) codes can be simply derived by taking pairs of bits from the binary input sequence and converting them into pulses with amplitudes -3, -1, +1 and +3 corresponding to the binary pairs 01, 00, 10 and 11 respectively. This procedure can be extended by taking n bits at a time and converting them into an m-ary signal where $m = 2^n$.

Another technique using multilevel signalling is the generalised partial response coding (Kretzmer (3)). The linear pseudo-ternary codes described earlier are the simplest forms of partial response coding. Higher order

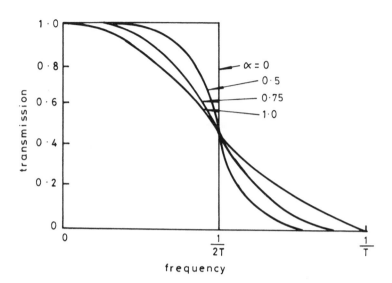

a   r.c.r.o. characteristics

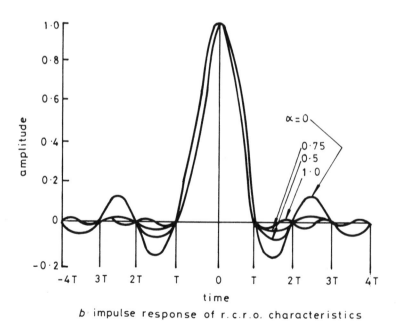

b impulse response of r.c.r.o. characteristics

Fig. 3.5   RAISED COSINE ROLL-OFF CHARACTERISTICS AND IMPULSE RESPONSE

partial response codes make use of an increasing number of amplitude levels to define the signal. The main disadvantage of the multilevel codes is that the higher number of levels makes them very vulnerable to interference from external sources such as crosstalk, impulsive noise and radio pickup.

3.3.2 Channel filtering

If pulses are transmitted over a band-limited channel they are inevitably distorted and dispersed in time. Thus, if a data sequence is represented by a stream of consecutive pulses, these pulses are bound to overlap on receipt, giving rise to intersymbol interference. Nyquist has shown (4), however, that if the channel frequency v gain characteristic has

odd symmetry about a cut-off frequency equal to half the pulse transmission rate, and has a linear phase characteristic, then the channel impulse response has zeros about its peak at intervals equal to the transmitted pulse spacing. Thus, if the received signal is sampled at the appropriate instants where the zeros preceding and following pulses occur, then it is possible to detect the transmitted sequence without intersymbol interference. A suitable response which can be closely approximated to in practise is that known as the 'raised cosine roll-off' characteristic. This is illustrated in Fig.3.5. The impulse responses corresponding to various roll-off factors are also shown. It will be seen that the oscillations in the pulse tails decrease as the excess bandwidth is increased, making the sample timing less critical. It is usual to provide the channel shaping by means of two identical filters, one at the transmitter and one at the receiver, such that the overall characteristic of filters plus channel gives the desired response. In this way the received signal-to-noise ratio is optimised.

3.4 EQUALISATION

Imperfections in the transmission channel characteristics will inevitably lead to departures from the ideal channel impulse response we have endeavoured to obtain for our base-band channel by careful design of the shaping filters. Instead of passing through zero at intervals of time T, the impulse response takes on values as shown in Fig.3.6.

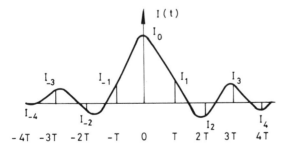

Fig. 3.6 IMPULSE RESPONSE HAVING INTERSYMBOL INTERFERENCE COMPONENTS

This gives rise to intersymbol interference. This intersymbol interference can be reduced by the use of a transversal equaliser (Lucky, Salz and Weldon (5)). The basic transversal equaliser consists of a delay-line tapped at intervals of time T. Each tap has associated with it a variable gain device as shown in Fig.3.7. The outputs from these multipliers are added together in a summing network to provide the equalised output. By suitably adjusting the gain settings it is possible to reduce the intersymbol interference. Enough delay-line taps have to be provided in order to be able to reduce the intersymbol interference to an acceptable level.

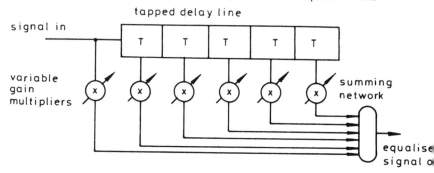

Fig 3.7   BASIC TRANSVERSAL EQUALISER

Normally the transmission circuit to be used, and hence the channel characteristics, are not known until the sender is ready to transmit data over the network. This means that equalisation has to be caried out each time a circuit is acquired for data transmission purposes. Sometimes the circuit characteristics are time-varying at a rate which allows significant changes to take place in the period of time the circuit is occupied for data transmission purposes. Under these conditions it is necessary to make the equalisation process adaptive. To automatically pre-set the equaliser, a test pattern is transmitted and the a priori knowledge of the pattern at the receiver is used to compute the impulse response from the received signal. From the impulse response so obtained it is then possible either to

calculate the tap coefficients directly or to use an iterative technique to successively increment the coefficients until an optimum setting is obtained. This mode of operation gives fast initial setting up but, if the channel characteristics are changing, it becomes necessary to retransmit the test pattern at intervals to allow for re-setting the coefficients.

An alternative mode of operation is that generally referred to as adaptive equalisation. Instead of using a test pattern the method is based on the assessment of the error between the actual received signal and the a posteriori estimate of the transmitted symbol at the receiver. The error signal obtained is correlated with the data stream to obtain estimates of coefficient setting errors. The coefficients are updated in accordance with these estimates to minimise the magnitude of the error signal. This strategy has the advantage that the setting-up procedure is a continuously adapting feed-back controlled process. However, if the unequalised signal is so badly distorted that a large proportion of erroneous decisions are made, the equaliser may not initally converge and equalisation will not be achieved. Under such conditions it is possible to start up in the pre-set mode, using an initial test pattern, and then to change to the adaptive mode when a reasonable degree of equalisation has been achieved.

Equalisation of modulated data signals is usually carried out after demodulation back into base-band. For quadrature amplitude and phase modulated signals, the demodulated signal can be resolved into two quadrature components. Each component can then be equalised separately. There will also be some cross-modulation components which can be dealt with in a similar fashion to the interfering components in the main signal. A schematic diagram is given in Fig.3.8.

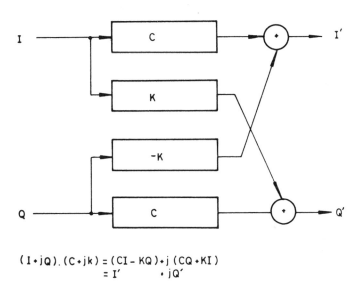

$$(I + jQ).(C + jk) = (CI - KQ) + j(CQ + KI)$$
$$= I' \qquad + jQ'$$

Fig 3.8 PHASE-QUADRATURE EQUALISER

Since both signal components are operated on by similar channel characteristics, it is often convenient to consider the two input signals as the real and imaginary parts of a complex signal and, likewise, the main and cross-modulation coefficients of the equaliser as the real and imaginary parts of an array of complex equaliser coefficients. Thus

$$(I + jQ).(C + jK) = I' + jQ' \text{ as shown in Fig 3.8}$$

3.5 DUPLEX OPERATION

In order to obtain duplex working over a single channel, some method of separating the forward and return signals is required. There are basically four methods that can be used.

3.5.1   Frequency separation

The forward and return signals are modulated onto different carriers so that the two signals fall into non-overlapping frequency bands. These can then be separated at the receivers by filtering before demodulation. This technique is mainly used on duplex voice-band modems and the V21 modem described earlier is a particular example of this technique.

3.5.2   Time Separation

The channel is used alternately to exchange groups of bits first in one direction and then in the other (Ithell and Jones (6)). Since a guard time is required between alternate transmissions to allow for propagation along the channel, it is desirable to send bursts of bits rather than single bits in each direction.   This mode of operation is therefore often referred to as burst mode or 'ping-pong' operation.

3.5.3   Hybrid Separation

The principle of hybrid separation has been used for many years to separate the forward and return signals in conventional telephony in order to convert from two-wire to four-wire transmission in the trunk network. In the case of digital transmission we are less concerned with transmission loss throught the hybrid and a simple Wheatstone bridge arrangement can be used.   Although the concept is basically simple, the decoupling between the forward and reverse channels is limited by the accuracy with which the input impedance of the line can be matched by the balancing network.   A digital processing structure can be used to generate inverse trans-hybrid characteristics to cancel the unwanted signal at the hybrid receiver output port.   Such a device has become known as an echo-canceller (Mueller (7)). The echo-canceller has a form very similar to the transversal equaliser and can, if necessary, be made adaptive.   Although the adaptive echo-canceller is complex, it is very effective in reducing cross-talk in duplex operation.

3.5.4   Code Separation

Coding techniques can be used in two ways to achieve duplex operation.   Firstly, different line codes can be selected for each direction of transmission such that the line spectra of the two signals occupy substantially non-overlapping frequency bands in the channel.   The signals can then be separated by filtering at the receiver.   Alternatively, orthogonal codes can be used, for example codes based on the Walsh functions, which can be separated by correlation at the receiver. Unfortunately the orthogonality of codes can be seriously affected by distortion caused by the channel characteristics, giving rise to cross-talk between the transmit and receive signals.

REFERENCES

1. Duc, N. Q., and Smith, B. M., 1977,
    'Line coding for digital data transmission',
    Aust. Telecom. Rev., 11, 14-27

2. Crosier, A., 1970, 'Introduction to pseudo-ternary transmission codes',
    IBM Journal Res. Dev., 14, 354-367.

3. Kretzmer, E. R., 1976, 'Generalisation of a technique for binary data
    communication', IEEE Trans. Comm., 14, 67-68

4. Nyquist, H., 1928, 'Certain topics in telegraph transmission theory',
    Trans. AIEE, 47, 617-644

5. Lucky, R. W., Salz, J., W. G. T., 1978, 'A proposal for the
    introduction of digital techniques into local distribution', Int. Seminar
    on Digital Communications, Zurich.

6. Mueller, K. H., 1976, 'A new digital echo canceller for two-wire full-
    duplex data transmission', IEEE Trans. Comm., 24, 956-962

Chapter 4

Data communications

Ian Paterson

4.1.   INTRODUCTION

When the user presses a key on the terminal, logic converts the action into an electrical representation of the selected character. This is sent to the remote computer via the communications channel. The channel may be just a few feet of cable or may consist of many components interlinked around the world.

Whatever transmission method is used, there must be a control process which accepts the data, possibly formats the characters into a message and copies the message to the application program. And handles error control. This is a job for a Protocol.

Where many remote terminals are in use, this control process imposes a heavy workload on the main computer so a dedicated communications processor is often used. This device is often called a Front End Processor or FEP, frequently a special purpose minicomputer with its own control programs and bulk storage.

Similarly, it makes systems sense to move functions like editing nearer or into the terminal and to handle terminals in groups to make better use of a shared communications channel. Products which do this are called Concentrators, Control Units or Multiplexers.

These more complex devices are still terminals as far as the communications channel is concerned. Where the channel uses telephone lines or a public network, the channel interface generally separates data processing and data communications, requiring standards to ensure interworking between the two disciplines.

4.2.   SERIAL TRANSMISSION

4.2.1.   Internal Representation of Data

Computer codes are a means of representing the familiar characters and symbols we write and speak in a form suitable for machine processing. Data characters in a computer system or terminal are represented by a group of bits (binary digits) each bit taking one of two binary states, 0 or 1, and using discrete electrical values. The group of bits is variously called a word, character or byte.

Since each bit can have two values, seven bits have 2^7 combinations or permit 128 different characters. Common business language needs about 100 symbols so seven bits have been chosen for many applications. Except where there is a special requirement, the great majority of terminals use a standard code to ensure widespread application. This is the I.S.O. 7-bit code.

The ISO code is promulgated by the International Standards Organisation. The United States version of the ISO code is the USA Standard Code for Information Interchange, usually abbreviated to USASCII or simply ASCII, and the user will find that most terminals available use this code. Of the 128 combinations 32 are reserved for control functions. In serial data transmission systems the ISO code is always sent as an 8 bit group, the additional bit being used as a check bit for error detection.

A current problem is; how to handle Word Processors? These have a larger character repertoire and need additional control characters not used by data communication, for example, underline and character fonts.

Bit Number			0 0 0	0 0 1	0 1 0	0 1 1	1 0 0	1 0 1	1 1 0	1 1 1
7 6 5	4 3 2 1		0	1	2	3	4	5	6	7
0 0 0	0 0 0 0	0	NUL	DLE	SP	0	@	P	`	p
0 0 0	0 0 0 1	1	SOH	DC1	!	1	A	Q	a	q
0 0 0	0 0 1 0	2	STX	DC2	"	2	B	R	b	r
0 0 0	0 0 1 1	3	ETX	DC3	#£	3	C	S	c	s
0 0 0	0 1 0 0	4	EOT	DC4	$	4	D	T	d	t
0 0 0	0 1 0 1	5	ENQ	NAK	%	5	E	U	e	u
0 0 0	0 1 1 0	6	ACK	SYN	&	6	F	V	f	v
0 0 0	0 1 1 1	7	BEL	ETB	'	7	G	W	g	w
0 0 0	1 0 0 0	8	BS	CAN	(	8	H	X	h	x
0 0 0	1 0 0 1	9	HT	EM	)	9	I	Y	i	y
0 0 0	1 0 1 0	10	LF	SUB	*	:	J	Z	j	z
0 0 0	1 0 1 1	11	VT	ESC	+	;	K	[	k	{
0 0 0	1 1 0 0	12	FF	FS	,	<	L	\	l	|
0 0 0	1 1 0 1	13	CR	GS	-	=	M	]	m	}
0 0 0	1 1 1 0	14	SO	RS	.	>	N	^	n	~
0 0 0	1 1 1 1	15	SI	US	/	?	O	_	o	DEL

EXAMPLE

Bits (P) 7 6 5 4 3 2 1
A = 1 1 0 0 0 0 0 1
i = 0 1 1 0 1 0 0 1

Notes

(P) = Parity bit

"A" with Odd Parity

"i" with even Parity

Figure 4.1   The ISO 7-bit code

Within the computer or terminal the data code is moved in parallel or all bits simultaneously, along 8 (or more) wires called a Highway or Bus. A short, external, parallel bus is usually available on most Microcomputers for printer attachment.

4.2.2.   Serial Representation of Data.

Serial transmission is used, almost without exception, for "out of building" applications. Less wires are needed, but additional terminal logic is required to convert the parallel internal representation to serial form for transmission. Serial transmission imposes strict limits on the data presentation; electrical values, duration, where the byte starts and ends and many other things need to be agreed between the two ends before successful operation is possible.

The method used is to define equal, short intervals of time as representing successive bits in the byte. Two possible conditions, termed "mark" and "space", are assigned specific voltage levels representing the 1 or 0. The byte is sent along the channel one bit at a time, usually sending the least significant bit first. The result is a series of voltage steps from 1 to 0 and back, a square waveform, being sent to the channel. This form of signal has limited use over the telephone wires, so is only usable for short distances.

Most computers and terminals operate using plus and minus 6 volts over about 50 feet. (see V24 below). Telegraph terminals use larger voltages for longer distances. Both these methods are known as bipolar or double current techniques. "Current Loop" terminals generate or interrupt a constant current around a loop. This is a limited distance technique but with high immunity to electrical interference. This method is single current working.

4.2.3.   The Interface.

The electrical values, meanings and timing are all collected together into a number of standards to ensure interworking between terminal and computer. The most common standard is CCITT V.24 and its American counterpart EIA RS232-C. Specific functions are associated with each pin on it's 25 pin connector, Data Terminal Ready (DTR) or Clear To Send (CTS), for example. The direction is from terminal, or DTE, towards network or DCE. The terminal usually has a cable, terminated with a plug, the DCE offers the matching socket. Ohms Law restricts the maximum bit rate to about 20,000 bits/sec and a maximum distance between devices of about 15 metres. For faster speeds than V24, CCITT V35 operates up to 72,000bps using a balanced driver/receiver pair for the important signals.

All the CCITT V series specifications are designed for use on the analogue PSTN. The newer X series of recommendations apply to the digital networks. Amongst the X series recommendations, X21 takes a new direction in specifying character sequences per pin on the connector. For example; if the control lines are off, then data on the transmit line will be accepted by the network as dialling information. Once connected, the control lines will be on and data on the transmit line is passed transparently by the network. This technique became necessary because V24 had run out of new pins for new functions. X21 means fewer pins on a lower-cost connector, future growth by using different character sequences and higher speeds and longer distances using V11, which has a fully electrically balanced transmitter/receiver pair.

The American RS449 was another attempt to address the shortcomings of V24/RS232, by using a larger connector. In fact two connectors are specified, a 37 pin and a 9 pin "D" types. This has not won much public favour, except that the electrical interface uses RS423 or 422, the latter being the equivalent of V11, with all the advantages above.

Another current change is to separate the interface into logical components; physical features separate from the method of use. X25 is probably the best current example of this, and is itself a member of a higher structure; the ISO Model for Open Systems Interconnecton. This "layers" communication between two end processes such that higher layers are insulated from lower ones. The top 3 layers of the 7-layer model are the subject of much Standards work. (See section 4.11)

4.2.4.   Start-stop or Asynchronous Transmission.

This mode of transmission permits characters to be sent one at a time, asynchronously with respect to each other. The transmitter uses a simple envelope technique which begins by preceding the data byte with a space or 0, with a mark, or 1 condition at the end of the byte. The leading space is always one bit duration, the trailing mark is usually one bit time but may be longer. The line remains at mark during idle periods.

There is always a transition from mark to space at the start of the first interval, used by the receiver to indicate that it must start sampling at the second interval. After this, the receiver continues to sample until the ninth interval whereupon it resets itself and waits until it sees another transition from mark to space. The first and tenth intervals are called respectively the "start bit" and "stop bit". They do not convey data but serve merely to provide an unambiguous indication that a byte is about to be transmitted or has just ended. The complete sequence is often referred to as a 10 unit envelope or 10 bit character.

To ensure correct operation, the time intervals used by transmitting and receiving terminals must be the same. The transmitting and receiving rates are usually expressed in "bits per second" (bps) and are controlled by internal clocks in the the two terminals. In the

Start-Stop method, the receiver clock is restarted afresh for each new character, stopping between characters. There may be a gap of any length between character transmissions.

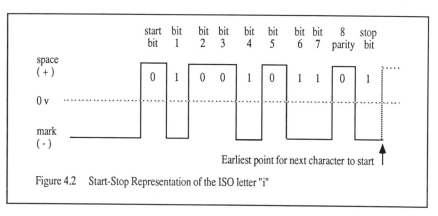

Figure 4.2   Start-Stop Representation of the ISO letter "i"

Asynchronous transmission is suited to terminals without storage and typically for interactive work, usually associated with Mini- and Microcomputers.

4.2.5.   Synchronous Transmission.

Although simple to implement, start-stop transmision is inefficient since ten bits are sent for every eight bits of data. Better use of the channel would result if large blocks of data bits were sent with an overall envelope. This would need a long term timing relationship to keep the two ends in synchronisation, since we may send many hundreds of bits. The timing relationship between sender and receiver is usually provided by the communications equipment (i.e. the modems). It is from this that the name originates; synchronous or isochronous transmission.

The data block is usually preceded and ended by specially defined characters. The preamble has a bit pattern which allows the receiver to "lock" it's bit and character clocks to the incoming data stream. This is necessary since all the information bytes follow immmediately with no interbyte markers. The postamble allows all bits to get into the receiver before anything else happens. The large block of data makes a fine "target" for errors so the block needs some form of error-correcting mechanism. This will need a protocol, or end-end agreement to decide what to do if an error is detected. Section 4.10 discusses Communication Protocols, but at this point it is worth noting that most synchronous operations need a protocol, and that a two-way channel is needed.

The technique requires more logic in the terminal which will almost certainly have some storage. The technique is usually associated with mainframe communication.

4.2.6   Character Rates

A point worthy of comparison is the information character rate of the above two transmission modes. For low volumes of data there is little real difference, but if a personal computer requests a screenful of data from a remote host then the choice of Start-Stop or Synchronous transmission may have quite different effects. A VDU screen is typically 1920 character in size. This is 19,200 bits of start-stop data (10 bits/character) or 15,360 bits of synchronous data (8 bits/character), an obvious difference in transmission and response times. An entirely opposite view is the effect on the transport network, which the user doesn't see: each 1000 bits of start-stop data is 100 characters to be buffered, queued, transmitted and possibly retransmitted in the event of errors. The same 1000 bits of synchronous data is 125 characters of buffer and frame space: buffer storage comes cheap but the manipulation of the blocks will take time, adding delay which the user does see.

The final choice is usually a compromise made on a wide number of criteria based on the overall requirements of the user population.

4.2.7.   Half and Full Duplex Transmission

A communication channel is described as full duplex if data can be transmitted in both directions at the same time. If data can be transmitted in either direction but only in one direction at any one time, the channel is said to be half-duplex. Both imply a two-way channel, one-way channels are sometimes used for data transmission, but these are not discussed here.

A full duplex channel may be constrained to operate half duplex by the overall system or target application. The process of changing the direction in which data can flow on a half duplex channel is called "turning the line round" and the time taken to do this is called the "turn-round time". Many terminal networks which do not actually require data to be transmitted both ways at the same time nevertheless use full duplex data channels, because the turn-around time of the channel would lead to unacceptably long response times from the computer. Another reason for avoiding half-duplex operation is that to turn the line round needs special functions in the computer software and in the terminal and computer interface hardware. Networks using low cost mini- and micro-computers and/or simple terminals usually use full duplex working.

The need to make future systems compatible with satellite communications channels which involve relatively long transmission delays, which need a protocol and which cannot be turned round quickly, means that most will use full duplex techniques.

4.3.   LINES FOR DATA TRANSMISSION

Over short distances, up to about 50 metres, the line has little impact on the digital signal from the terminal and hence provides a simple digital channel. The most commonly used medium for transmission both nationally and internationally is the telephone system. Links may be established by dialling on the PSTN or by using private circuits leased from the the Carriers.

The leased lines were designed for voice and are specially equalised or "conditioned" for data, which improves the quality to some extent, however little can be done to improve the channel obtained over the PSTN. In this case the data communications equipment must be able to tolerate the variability and correct induced errors. As well as conditioning the lines to give a constant performance, the data user needs a device to convert the digital data into a form suitable for the analogue channel, the modem.

Short distance lines within an exchange area could use Baseband modems, limited distance modems or line drivers to produce the digital channel. The higher the speed of the modem, the more of the available bandwidth on the line it needs, bandwidth which is restricted due to its use for ordinary voice telephony.

Where private lines may be installed, within a building for instance, this bandwidth restriction may be removed, providing a channel with a capacity of many millions of bits per second. I am of course talking about Local Area Networks or LAN's. Two techniques are in common use; Broadband, which uses frequency separation, and Baseband which relies on various forms of arbitration to provide access on demand, either direct competition or by using a "permission to send" mechanism (CSMA or Token system).

One circuit between two offices may use a combination of many media; Frequency division multiplexing on older concentrators, Time division multiplexing on digital trunks, possibly over Fibre optic links and even via satellite to another country, with the humble copper pair in the local loop. For the first time in Data Transmission history, the availability of digital services means that the long distance channel is starting to offer bandwidth which does not restrict the network design, even though the raw bit rate is beyond the ability of most computers to use directly.

4.4.   DATA NETWORK DESIGN TECHNIQUES

The lines used for data communications are so costly compared to the processing equipment that most of the techniques used in terminal networks optimise performance against the tariffs. The methods employed aim to move the maximum amount of data with the smallest number of lines. Some networks are installed regardless of this parameter due to other operational requirements becoming more important. For instance, rapid response is related to the number of terminals sharing the line as well as the line discipline used, the volume of data and the turnaround time. High availability may mean a very low terminal to line ratio.

4.4.1.   The Point to Point Network

In the point-point network, each terminal is connected to the computer by a direct line (see figure 4.3). This always creates a star network with all lines radiating out from the computer. This is an ideal network in many ways; terminal and computer have immediate access to each other with minimum effort and delay. The software is simpler since each line goes only to a defined terminal. Response is very good because the full speed and capacity of the line is dedicated, however, few very large networks of this type exist due to the sheer size of the connection and accomodation problems at the host site.

Point-to-point is highly efficient between computers or between dedicated processes, but human thinking and action times usually make poor utilisation of the circuit.

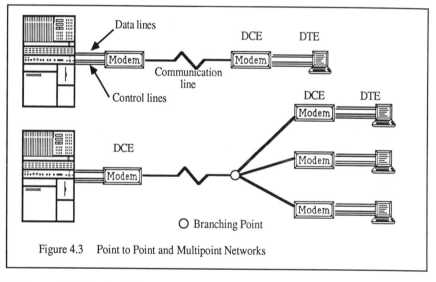

Figure 4.3   Point to Point and Multipoint Networks

4.4.2.   Multipoint Networks

An obvious way to improve this utilisation level is to share the circuit between several terminals. This may be implemented by "clustering" the terminals at a point or by "multidropping" terminals at different points along the route. Both systems are called Multipoint networks with the junction points called Branching Panels. (See figure 4.3).

The economics now look much better, however, to avoid chaos, the computer software must determine when each terminal is allowed to transmit and must inform each terminal when to receive. This technique of scanning and selecting is known as "polling".

Polling has the advantage that each terminal enjoys full access to the line when selected, but has the disadvantage that data must be buffered when the terminal is not selected. This also means delay in response when one terminal is blocked by another one. Simple mathematics can be used to calculate the maximum number of terminals on a line given its speed, the message sizes, and turn around time; but in practice typically up to five terminals are used before the users complain of poor response.

The disadvantages are as "bad" as the advantages are "good". If the line fails how many terminals will be out of service? The terminal must be more complex, intelligent and hence expensive. Special software (and computer time) is involved in the line protocol and how it controls each terminal. And what to do if one terminal insists on "speaking" all the time despite not being polled? (a "streaming" terminal).

4.4.3.   Multiplexing

Multiplexers enable a number of low speed terminals to share a common higher speed communications line. Each terminal and port behave as though they were connected by a point-to-point circuit, having exclusive use of their own share of the bandwidth without being aware that multiplexers are being used. The terminal is attached to the "low speed" side of the multiplexer, the "high speed" side uses the communications line. Multiplexers are used in pairs.

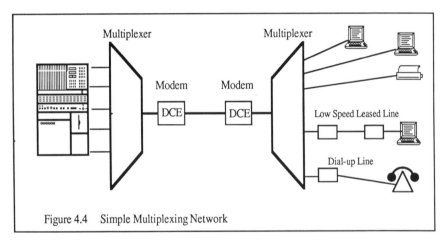

Figure 4.4   Simple Multiplexing Network

Since multiplexing doesn't need complex terminals, it is a technique used more frequently for asynchronous rather than for synchronous terminals. There are three basic techniques; frequency division, time division and statistical multiplexing.

4.4.3.1.   Frequency Division Multiplexing

This is the oldest technique, used as the basis of the voice telephone network, and is known as FDM. It takes advantage of the fact that the bandwidth of the line is significantly greater than that of the low-speed channels. Using different centre frequencies several channels share the line using an analogue filtering technique, like the familiar radio bands.

Channels have "guard bands" of wasted bandwidth to prevent adjacent channel interference or crosstalk. Bandwidth is limited by the high speed circuit. Changes are difficult to implement since it may involve "reprogramming" a number of other channels. FDM is unsuited to handling synchronous terminals due to the difficulty in deriving and synchronising the clocks. FDM has not found it's way into the commercial arena until very recently, where it is used most successfully to provide Data over Voice links, usually on in-house circuits. FDM is of course the basis of Broadband Local Area Networks.

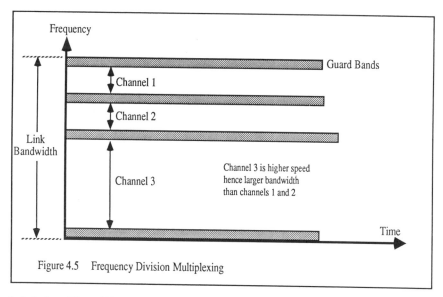

Figure 4.5   Frequency Division Multiplexing

4.4.3.2   Time Division Multiplexing.

Time Division Multiplexing, or TDM, is a digital technique employing the interleaving of bits (bit TDM) or bytes (byte TDM) from the low speed channels onto a higher speed composite link. The composite channel is "sliced" into short time periods which are assigned to the low speed channels in some strict order. These "time slots" are assembled into a "frame", and the frame structure is sent repeatedly.

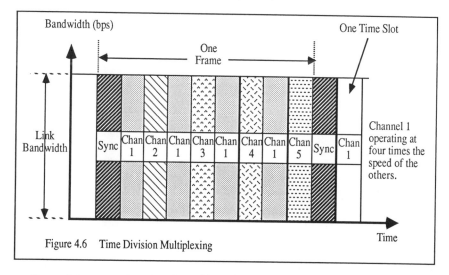

Figure 4.6   Time Division Multiplexing

Two points are worth noting about this technique. Firstly the multiplexers must be in synchronism with each other, otherwise the data would go to the wrong channel when demultiplexing was performed. This is achieved by the regular insertion of synchronising bits or bytes which separate the frames into fixed length blocks. And secondly, since the frame repetition rate is fixed (fixed length frames over a fixed speed channel), then the number of time slots for a given low speed channel determines it's maximum bit rate.

All these points are illustrated in figure 4.6, which shows channel 1 operating at four times the speed of the other channels.

TDM relies on speed changing using buffers; characters (or bits) arrive from the low speed channels and are held until "their" timeslot occurs. In general, Character interleaved TDM is more widely used, because of it's efficiency in multiplexing asynchronous terminals. Bit TDM adds less delay and is much more efficient at handling synchronous channels. Bit TDM is more usual at higher composite speeds of 64Kbps and above, as seen on the British Telecom Megastream service.

4.4.3.3.   Statistical Multiplexing

Simple TDM sends fixed length frames containing time-slots for each channel. The frame must be sent continually to maintain end to end synchronisation between the multiplexers, also the time-slots must be "padded" with idle characters (or bits) when the corresponding low-speed channel is silent. Errors may cause resynchronisation between multiplexers and the effects of both errors and resynchronisation have to be handled by the external equipment.

With the advent of the low-cost microprocessor, intelligence can be added to the multiplexer. Variable length frames need only carry data from the currently active channels hence more channels may be handled per frame. The frame may have a block checksum added to enable error checking. The frames are usually numbered so that retransmissions may be made to provide error correction. Numbering also provides a capability for overflow protection by limiting the maximum number of frames acknowledged by the receiver.

Since time-slots are used proportional to channel activity then a much lower-speed composite link may be appropriate. Alternatively, the sum of all the channel inputs may be many times the composite link rate since we are relying on the statistical probability that not all the channels will be active at the same time. This potential "overload" situation is called the compaction ratio, and relies on buffering to prevent data loss.

This technique is called Dynamic Bandwidth Allocation, or more commonly statistical multiplexing.

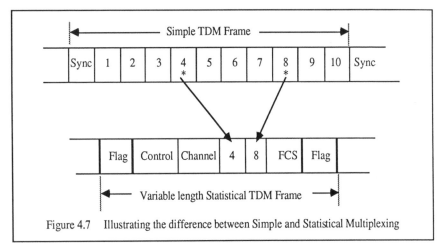

Figure 4.7   Illustrating the difference between Simple and Statistical Multiplexing

In the simple frame, time slots 4 and 8 (indicated by an asterisk) are the only active channels and "Sync" indicates the synchronisation time slot. In the statistical frame, Flag performs the same function as "Sync", "Control" enables frame identification for retransmision, "Channel" represents a method for knowing which channels are active and FCS is a frame checksum for error detection.

4.4.3.4. Networking properties of Multiplexers

Since TDM is a digital technique, it is a relatively simple task to demultiplex the frame and, instead of sending the data to the low speed channel, to remultiplex it with other data onto another composite link. This is sometimes called "Onward linking", "Mid-pointing" or "Space switching" and occurs at "Switching nodes" or "Cross-connect sites". This capability means that the multiplexer has multiple composite links, enabling star and mesh networks to be created. This basic capability may be extended where the outbound "composite link" may be a protocol converter, for example to a Public Packet network.

4.4.4. Optimisation

The ultimate objective of the the network is performance. But performance needs to be defined according to objectives; delay, response, cost, reliability. I know of no general rule to optimise a network, and since the exercise provides an intellectual challenge, you will read many articles and hear many strongly held views! So here are my suggestions.

The design usually begins from the geographic point of view. The terminal types and quantities and the computer centres will be linked together using a concept of "corporate correctness". By this I mean that most organisations install a network which reflects their corporate mode of operation; strong, central control produces a star network, autonomous regional control yields a linked mesh and so on.

This is followed by a review of what the network will carry. What constitutes a message? What is the frequency of arrival and the route to the destination? This yields capacity requirements and suggests techniques, modems, multiplexers and so on.

And then the hard part; features versus benefits. The network design will reach some point and will then become a compromise, except perhaps reliability where some networks are designed regardless of cost due to the commitment of the organisation to use it.

Optimisation also means recognising both your starting point (why you made certain decisions) and where you are going (the eventual aim of the network), and then reviewing how each step or phase of the implementation is taking you towards that goal. Optimisation really never stops; if your network appears to be static it's probably because someone else has plans to replace it!

And don't forget that any "good" network, from the user's point of view, will generate it's own traffic, just look at what happens when a new motorway is opened!

4.5. CIRCUIT SWITCHING.

The networked multiplexers above are an example of Circuit switching. Circuit switching is most familiar to us through the PSTN. This provides a two-way transparent path, without storage and with minimum end-end delay. Physical resources in terms of time, space or frequency are dedicated to the call. A bit stream entering a circuit switched channel will re-emerge at the other end with an identical time pattern delayed by a constant amount. And although the channel may have been derived from multiplexing, each circuit operates entirely independently of the rest. There is no flow control, both ends must operate at the same speed.

The design of a circuit switch is based on a "call", statistics of the call define the required performance; calling rate, duration, frequency. The perfect circuit switched network is a star, with all circuits coming into the switch. In real life this would be impossible, since this implies a massive termination problem. In practice, the circuits are "concentrated" before switching. The concentrator is not a true switch in that it cannot complete a call locally, merely connects active subscriber lines to a smaller number of circuits to the switch, usually using TDM. The concentrator relies on "grade of service", implying a knowledge of the number of required circuits to the switch, ie less circuits than subscribers, hence blocking or congestion can occur.

Since the switch connects a call on an incoming physical trunk to a circuit on another outgoing trunk, it is often known as a "space" switch, thus with concentration such a switch is dubbed a TST or "Time-Space-Time" system.

The circuit switch has substantial call set-up delay, since a complete end-end path is needed before transmission can begin. Network resources are dedicated on an "exclusive basis" ie one line - one call. More recent circuit switch designs with fast call set-up use a separate circuit or even network for control signals (see ISDN and 2B+D channels).

A (digital) circuit switch deals with a binary bit stream at a pre-determined speed, the common link between the two ends being bits/sec. The circuit switch is very appropriate for the telephone network due to the uniform bandwidth and grade of service, less so for data with it's wide range of speeds and different traffic pattern.

4.6. MESSAGE SWITCHING

The whole purpose of the Message Switch is to handle a whole message rather than establish a channel between the parties. Delay is not as important as integrity as there will be no delivery confirmation. Store and Forward Message Switching relies on always being able to accept the entire message whenever it appears, often serving a very large number of lines. The message is then delivered once the circuit to the destination is free. With a large variation in traffic load, excess traffic will be formed into buffer queues for the outgoing lines. Large peaks will cause the switch to place the message onto longer-term backing store, until capacity becomes available. In fact, most messages are stored for later retrieval anyway, since the technique lends itself to Government, Military or Business where there is a need for written records. (The method is sometimes called "Record Communication").

Message switches provide a high degree of checking and logging to provide safe delivery, being primarily a "writer to reader" type of transport service rather than real-time, thus needing a powerful processor with large amounts of storage. However, these provide the ability to handle incompatible terminal ends by providing code, speed and protocol conversion.

Whole messages are handled, therefore long ones will cause variable delivery delay, however the switch is non-blocking, and can always accept an incoming message. Also, the method leads to highly efficient use of network resources, with high utilisation on long-haul circuits. Store and Forward Message Switching does have the disadvantage that the message is always kept whole, and could be a privacy/security risk. But this technique is ideal for data devices which have a very high peak to average usage level.

4.7. PACKET SWITCHING

Packet switching is designed as a wide-area, data only service featuring error-correction, security, integrity of links using rerouting, speed conversion, flow control and multiplexing. Usually known by it's CCITT standard, X25, it combines the best features of Circuit and Message switching by splitting up the data into blocks for switching and transmission. Since the traffic is likely to be "bursty" in nature, X25 describes a multiplexing method based on "virtual circuits", many of which may be interleaved on to a single physical circuit. Packet switching allocates resources on demand unless contention occurs when queueing is used, or if congestion occurs when flow-control is invoked.

End-end connections are made using Virtual circuits, so called since no real circuit is dedicated to the call. The virtual circuit is set up for the call, determining the path for all packets which follow. The virtual circuit number is an "alias" for a particular call, Packet Switching Exchanges (PSEs) provide an address translation between the two ends.

The PSE can only switch packets, hence all attached equipment must "speak" packets. Non-packet devices will need a protocol converter; a Packet Assembler/Disassembler or

PAD. The asynchronous PAD is a parametric device, designed to handle a wide variety of different terminal types, and is described in CCITT X3, X28, X29 standards. All packet devices use X121 for their addressing scheme.

X25 describes a protocol with three layers or levels. At the bottom is the physical layer which deals with the electrical interface and the plug and socket. It specifies a synchronous, full-duplex path for data bits. Level 2 provides an error-free point to point link between the DTE and DCE, using HDLC (see below). This "wraps up" the user packets in control information for transmission.

Level three is the packet level, defining the format of the packets and a call handling process. Two types of channel are available; Switched Virtual Circuits and Permanent Virtual Channels, SVC and PVC for short. The SVC uses Call Request packets to set up switched calls to any permissible destination, the PVC is the equivalent of a private, leased line. Level three provides for massive future growth; a maximum of 4095 logical channels on a single physical link, a flow-control process allowing up to 128 blocks of upto 2048 bytes of traffic per virtual channel. Modern processors are too small to handle this, so the performance is limited by the Access Node to the network.

Packet Switches provide the ideal compromise between the low delay of the circuit switch and the high efficiency of the message switch, without the disadvantages of either. The method may be described as "hold and forward", where a copy of the data is held for retransmission but not for long term storage. The switch cannot guarantee delay nor bandwidth, however it gives rapid call set-up and efficient use of circuits, whilst giving good response due the small, regular size of the packets.

Packet switching is ideally suited to data transmission as it is non-transparent, sensitive to the peculiarities of the terminal; speed, code, flow-control, for example, and can be used in interactive connections. Probably the major advantage is that X25 is a real, industry-wide standard, with the unifying effects that creates.

4.8.   SWITCHING COMPARISONS

Data from a computer is not a continuous stream, rather a sequence of messages. The packet switch was developed where short messages or packets are to be handled by the network. The packet switch involves storage but operates at a speed adapted to suit the terminal. Packet switching is a good choice for interactive work.

The circuit switch is ideal for bulk traffic, where the clear channel and constant bit rate will move the maximum data in the minimum time. Also, the larger call set-up times don't matter where the circuit is probably established by the user rather than machine. And the low fixed delay makes this an obvious choice for voice and any control applications.

The Message switch is used in non-real-time situations where guaranteed delivery is required. It is also suited to broadcast situations. The Message Switch is most useful where people are involved since a new message will always be accepted.

4.9.   FLOW CONTROL

Information flow control has two purposes; to synchronise two remote processes and to prevent congestion. The former is related to capacity, the ability of a device to accept information. For example, the short-term "lumpiness" of flow to a printer is usually due to the need for speed conversion between the computer port and the mechanical printing mechanism. Flow control per se is an end to end process where the network needs to transfer the restiction back to the data source.

Congestion on the other hand, is a more localised phenomenon and is to be avoided at all costs, as the end result is usually a complete lock-up where no traffic is able to flow. Congestion avoidance is more necessary in non-transparent networks where queues may form, since queues mean delay. Avoidance is usually a form of self-protection, usually resulting in the network unilaterally discarding data to free buffer storage. Congestion avoidance is of particular importance with interconnected networks, where some form of address or protocol translation has to take place.

Flow control also ensures fairness; preventing a long transmission from monopolising the channel and shutting down the network to other users.

But most of all Flow Control is needed to prevent loss of information due to errors and to ensure that information is sent when the communications channel is free and the appropriate terminal is ready to receive.

There are a number of error handling techniques, data feedback, decision feedback and forward error correction, but first detect the error!

4.9.1. Error Detection and Correction.

Looking at error control first, the simplest method is Information feedback where the character or message is echoed back to the sender. This is often called "echoplex". The user resubmits any erroneous data.

All other error detection systems use some form of additional check bits added to the data before transmission. The simplest check involves the addition of a "parity" bit to the seven bit ISO character such that there is an even number of 1's in the character (even parity) or an odd number of 1's (odd parity). This enables the receiver to check for bit errors where a 1 becomes a 0 or vice versa. Asynchronous transmission typically uses even parity, synchronous transmission uses odd parity. However, the parity system is "blind" to even numbers of errors.

In an attempt to overcome this problem, a second parity system is sometimes used. This is a "lengthways parity" through the whole block of data ie. the same bit of every character is counted. This is repeated for every bit, including the lateral or original parity bits. This parity "word" is called an LRC or Longitudinal Redundancy Check.

This system is also vulnerable to undetectable errors, especially bursts of noise, so a new technique is needed. This is the CRC (Cyclic Redundancy Check). The whole block of data is treated as a single binary value and is divided by a predefined number, the remainder being appended to the block as the CRC character(s). CRC checking relies on the fact that each data bit is used many times thus the chance of an undetectable error is small. All errors involving odd numbers of bits and error patterns shorter than the check character will be detected by the CRC system.

With all parity or CRC systems, decision feedback is now used where the receiver signals a positive or negative acknowledgement (ACK or NAK) back to the transmitter. Recovery usually needs retransmission.

Forward Error Correction (FEC) employs such a large number of additional bits that the receiver can reconstitute the data, identifying and correcting any errors. This method is common in military and spacecraft systems, where there may not be the opportunity to repeat a message. Hamming codes are an example of this system.

Error correction is usually handled by the communications protocol and is done in real-time; it would be foolish to transmit for long periods if an error needs retransmission as the transmitter would have to store all the data "just in case".

4.10. COMMUNICATION PROTOCOLS

The term protocol is used to describe a set of rules governing information flow.The simplest protocol is that used for start-stop terminals. The V24 interface has separate data and control lines, the latter being used to determine who shall send and who shall listen. This is called "Out-of-Band" flow control since the signalling doesn't involve the data circuit.

Alternatively, a modern printer with an electronic data buffer may be able to accept data from the line at 960 characters/sec, whilst the mechanical printing process cannot operate faster than 180 cps. When the buffer is full, the printer sends a character to tell the computer to pause whilst printing takes place, and later, when to restart. This is "In-band" flow control, typically using the ISO characters DC1/DC3 or XON/XOFF.

But the term "protocol" really applies to block-mode operation; all synchronous and many asynchronous devices are now using error-protecting block-mode protocols.

4.10.1. Simple Block-mode protocols

In block-mode, the data is gathered together into a block to which control and check data are added before sending. In a multipoint system, part of the control "overhead" would be a separate "conversation" to select the terminal next allowed to transmit. This selection or scanning process is called polling. This prevents confusion on the line, only the desired terminal is selected, and the computer can ensure that nothing is sent unless the terminal is ready and working correctly.

Once selected, a terminal may send a block of data but must await acknowledgement before continuing. The receiving device sends a short supervisory block to indicate error-free transmission or that an error occurred and retransmission is required. This protocol also guards to a high degree against total loss of data; if the terminal receives nothing or becomes inoperative it will not respond.

This type of system is called "Idle-RQ" (Idle awaiting Repeat reQuest) or called a "half duplex" protocol because after each block is transmitted the sending terminal waits for a response from the other end. On satellite links, which have a long transmission delay time, the use of these protocols would waste a lot of channel time while terminals wait for the responses. To overcome this problem a full-duplex protocol is used.

4.10.2. ARQ Protocols.

The sending end transmits block after block without delay, and only takes action when it receives notification of an error. By this time, of course, several more blocks will have been sent. The system must therefore number the blocks individually and must keep blocks in storage for the time necessary for an error notification to arrive. The error notification must indicate which block is in error. This type of system is called ARQ (Automatic Repeat reQuest). It is usual to permit the transmitter to send a pre-set maximum number of blocks - "a window" - before stopping until at least one block has been acknowledged. In addition, multiple acknowledgements are possible, for example one end sending a small number of very large blocks will acknowledge several blocks at a time if the other end is sending a large number of small blocks.

Although ARQ protocols are usually associated with synchronous operation, there are several examples of start-stop versions which use the technique purely for error-correction. For example, the British Telecom E-PAD is one of several ways to provide an error-free link from a terminal to a Public X25 PAD.

4.10.3. H.D.L.C.

HDLC, which stands for High level Data-Link Control, is an ARQ protocol. It is a full-duplex protocol which provides an error-free point-point link. HDLC is used as level 2 of the Packet Switching standards.

The data blocks are called "frames", each separated from the next by one or more "Flag" characters. Immediately following the Flag is an Address field, selecting DTE or DCE. Following this is the Control field which contains commands, responses and sequence numbers of the blocks. Data follows this in the Information field, followed by the two byte CRC. Then the block is terminated by another Flag.

One Flag may end one block and start the next. The Flag is a unique bit pattern (01111110), to delimit the complete block. The use of Flags allows the block length to be variable, but means that the pattern must not be allowed to occur within the rest of the block. To achieve this, a technique called "bit stuffing" is used. This deliberately adds an extra 1 bit to the data stream whenever a 0 followed by five 1s is seen. Similar logic in the receiver performs a compatible "bit deletion" process on the data block. Due to this bit-level sensitivity, HDLC is known as a Bit Oriented Protocol, the examples in section 4.10.1 being Byte Oriented Protocols.

Within the Control field are the block sequence numbers, $N(s)$ and $N(r)$. The transmitter sets $N(s)$ as the number of this frame being sent, and $N(r)$ as the next frame number it expects to receive. If several frames have been successfully received then a single N(r) may be used to acknowledge them all. The sequence numbers are usually up to 3 bits long, ie values 0-7, but for satellite circuits may be up to 7 bits long (0-127).

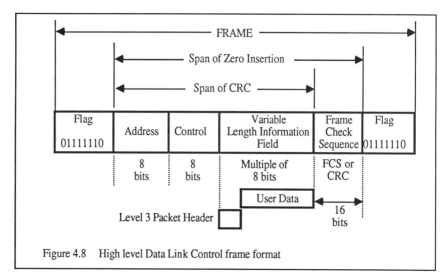

Figure 4.8   High level Data Link Control frame format

HDLC includes a comprehensive but surprisingly small group of commands and responses for complete link control. For example, if a receiver is very busy or has little storage a "receiver not ready" (RNR) frame tells the transmitter to hold off, "receiver ready" (RR) permits continuation. If a block is received with errors, the CRC check will fail and the block is usually discarded. The next correct block will have the wrong sequence number and causes a "reject" (REJ) frame to be sent requesting retransmission of block N(r). This also acknowledges blocks up to N(r-1).

Obviously, both ends must operate to the same protocol, and this is where standards are important.

4.11.   INTERFACE STANDARDS

Recommendations and standards have been produced by the CCITT (International Telegraph and Telephone Consultative Committee) and by ISO (the International Standards Organisation) and within the USA by EIA (Electronics Industries Association) and by ANSI (American National Standards Institute).

Some of the more common standards are;

V.24/RS232-C	Most widely used. Designed for terminal to modem but now used for almost any terminal application. Uses V.28 for electrical specification and ISO 2110 for plug and socket.
V.35	High speed version of above, 48, 56, 64, 72Kbps. Uses a combination of V28 for control lines and balanced send/receive, and ISO 2593 for the connector.
X.21	Usually used in leased line manner, rather than the auto-dial feature Uses V.11 for electrical interface and ISO 4903 for the connector.
V.11/RS422	Electrical specification for balanced, high speed, long distance line drive capability.
V.10/RS423	Electrical specification for interworking with V.24 or V.11.
RS 449	Less popular American attempt to replace V.24/RS232-C.

X.3, 28, 29   "Triple X" standards for a PAD, converter for async and X.25
                X.3 for terminal profile.
                X.28 is the command language, async to PAD.
                X.29 is for command interchange, PAD to remote X.25 DTE.

X.25   Interface between DTE and DCE in Packet Mode

Without standards, users face limitations when creating, modifying or linking networks. Once, a network was supplied by the computer vendor, for complete systems compatability. Today, few vendors can satisfy all the needs and single vendor systems are less the norm. The user gets a wider range of products from which to choose. Freedom of choice also fosters competition which usually leads to product improvement. This is where the ISO model for Open Systems Interconnection comes into it's own. Standards are also "free consultancy" to manufacturers; simplifying Research and Development, manufacturing techniques, components and maintenance. This also gives the supplier a bigger market.

4.12.   SUMMARY

We have taken a brief look at wide area data networks superimposed on the PSTN. There is an underlying need for standards from end to end; from the character code in the terminal, across the interface into the wide area network, how the network switches and makes routing decisions, and how to keep both ends logically "tied together".

The network is seen by the user as a delay, the designer must have created a combination of elements such that the delay is acceptable. The elements include multiplexers, concentrators, control units, modems, lines, Public networks, Front end processors and all the hidden software (and hardware) needed to keep the system operational and to diagnose faults when they occur.

In the early days of networks, the cost per bit decreased as you went from low- to higher-speed lines. This "delta cost" provided the opportunity and incentive for multiplexing and multidropping, ie line sharing. But with digital circuits, it costs no more to instal a 2400bps circuit than a 64,000bps circuit. This will squeeze the price for add-on devices to improve the utilisation.

The longer term effects indicate a sharing of facilities, especially voice and data. Intelligent terminals will increase, the desktop micro becoming more common. But the effects on the network will be far-reaching; many short calls, set up by machines with immediate retry capability, will wreak havok with a PSTN expecting fewer calls per second, longer holding times and longer hold-off after blocking. And on Packet networks, loading a file from mainframe to micro will cause flow control and capacity problems. This diversity will cause a need for transmission capacity switching, moving from synchronous TDM with low delay, clear channels towards Packet Switching and store and forward Message Switching. The increase in single-user Personal Computers will stimulate the move to Message Switching, as well as causing a move from the dedicated telex machine towards X400, again helping with the high peak to average traffic levels. And digital networks should provide improved bandwidth and performance.

The other major factors will be security and privacy. As we move towards an integrated common-services network with a diversity of information form, we will want to control the communication link whilst expecting protection from the unauthorised caller.

REFERENCES

V Series Recommendations   CCITT Yellow Book Vol. VIII.1
International Telecommunications Union, Geneva

X Series Recommendations   CCITT Yellow Book Vol. VIII.2
International Telecommunications Union, Geneva

Communications Networks for Computers   Davies D.W.   Barber D.L.A.
John Wiley and Sons

Computer Networks and their Protocols   Davies, Barber, Price, Solomonides
John Wiley and Sons

Handbook of Data Communications
NCC Publications

Management of Data Communication
Datapro Publications, Geneva

Chapter 5

Traffic theory

Bob Whorwood

5.1 INTRODUCTION

The importance to the telecommunications engineer of traffic theory can hardly be overstated. When the telephone system is installed, it is far too late to realise expensive mistakes have been made. Calculations involving traffic theory are necessary at the planning stage. Practical users of traffic theory do not need to understand all the 'fine points' nor carry out complex calculations. They use traffic tables which encapsulate traffic theory and thus are using it indirectly.

For some this may seem sufficient but it is really not so. From a personal point of view there is not much 'job satisfaction' in just reading numbers off a table when one has no idea what they really mean. More practically, we need to have a feel for the consequences of deviating a little from the table values; it may, or it may not, matter very much. We need to know something of the assumptions which are implicit in the traffic tables we use. Maybe those assumptions do not quite fit our particular situation. For example, large groups of trunks are more efficient than small ones from a traffic carrying point of view but less capable of coping with overload. Why? A feel for traffic theory helps to understand such matters even if one passes over to a specialist the job of dealing with traffic problems.

Traffic theory was developed for telephone systems as they have been around a very long time, but it is important to remember that the theory is very general, and applies equally well to many other situations which are analogous to telephone systems. The theory may be applied to the provision of chairs in a barber's shop, provision of service in a cafeteria, or lanes in a motorway, and so on. Accordingly it will apply to forms of communication other than speech although some of the assumptions will differ from application to application.

The common ground between these diverse areas of activity

is that they all depend in a similar way on the uncertainty of events happening. The individual requiring service will know he wants to make a telephone call, have his hair cut or drive down the motorway. The provider of the service does not know who is about to demand that service next, but he is far from ignorant. He knows that somebody out of the population of potential users probably will, and from past experience, the average number who are likely to do so. The system, whatever it is, must be able to cope in some degree with the demand. How well, or how inadequately, the demand will be met is the question that traffic theory addresses.

Traffic studies are therefore based upon studies of probability theory. Usually, such matters are explained in the literature in terms of the probability of drawing (say) a King of Hearts and King of Clubs from a pack of playing cards. Another common illustration is based upon drawing certain coloured balls from an urn. It is a pity that writers on the subject do not employ more useful illustrations of the subject, such as the probability of a call arriving at a telephone exchange, as there is no difference in the basic ideas.

5.2 WHAT IS TELETRAFFIC?

Perhaps the best way to get a feel for the topic, if it is still possible to find one, is to spend a while looking at a traditional manual board. A few racks of Strowger switches may be a less attractive alternative. It is possible to see from the appropriate lamp displays (or seized Strowger switches) the three main components of teletraffic, namely CALLS IN PROGRESS, CALL ARRIVALS and CALL HOLDING TIMES.

If one starts counting the number of calls in progress, it is obvious that, by the time counting is finished, some of the circuits which were free, and hence not counted, will have become busy and some that have been counted will have become free during the time counting is in progress. With traffic theory this does not matter: it is assumed that the system is in statistical equilibrium which, in simple terms, means that the counting 'errors' made one way cancel out those made the other. This assumption may be avoided by taking a photograph of the board every two (say) minutes so that effectively one can thus scan all the circuits at such an interval. Then, if this is done 30 times over an hour, the average occupancy over the hour can be obtained. Naturally, in practice, this sort of thing would be done electronically but, here, the object is to get a feel for the subject.

Traffic sudies are concerned with when the demand for service is heavy. Clearly if demand is met when it is heavy, a system will cope with a light load. Traffic varies throughout the day and also from day to day. Often it is busiest mid-morning when people have dealt with their mail and the mail has caused there to be a need to ring others up. Conversely, Friday afternoons are often a very slack period on the network, probably because of people wanting to finish work a little earlier to get away for the week-end or at least 'leave it until Monday'. Seasons cause variations also. All telephone administrations attempt, by their call charging policies, to transfer heavy demands for service to periods when demand is light. Another variation is the exceptional and occasional day such as Mother's Day or other festive occasions or, maybe, when there is broadcast a television program involving 'Phoning In'. Do we cater for such events or not?.

Decisions upon when to take the busiest period, and how important it is to cater for it, will vary from network to network. Private networks differ enormously one from the other, in this regard. A network used by a manufacturing industry will carry little traffic during industrial holiday periods but the corresponding networks used by the travel industry will possibly experience peak demand.

Thus there are at least two approaches to the problem of choosing the busy period to be catered for. One is extreme value engineering which is an approach to be taken when, whatever the cost, the consequences of not meeting extreme demand are too disastrous to contemplate. A non-traffic example might be the building of a sea wall which must withstand any maximum water pressure considered remotely possible. Maybe financial institutions could provide a similar example with traffic, where failure to communicate could cost millions of pounds.

The other, more common approach, is to consider a normal and frequently occurring busy period. If measurements are taken during the busiest part of the day for (say) one hour, this is a period long enough to cancel out short-term peaks and troughs over that hour. The period of one hour is long enough for this purpose, but not so long as to be affected by factors such as lunchtimes etc. Perhaps half an hour would suffice. If this is done every day for one week each month and an average taken over several months a typical, but not extreme, busy hour loading is determined. To take account of maximum loads, a percentage overload is considered, this being typically 20%.

5.3 TYPES OF SYSTEM

There are three basic types of switching system, LOST CALL, DELAY and OVERFLOW. With lost call systems, a call not set up immediately, through lack of equipment, is lost. Until very recently such was universally the case and remains largely, if decreasingly, so on the public network. Delay systems are where calls are put into a queue, and dealt with later according to some strategy such as order of arrival,

random or last come first served.   Overflow is where an
alternative route is provided if the direct routes are all
occupied.

5.4 LOSS SYSTEMS

   The discussion below is confined to simple loss systems.

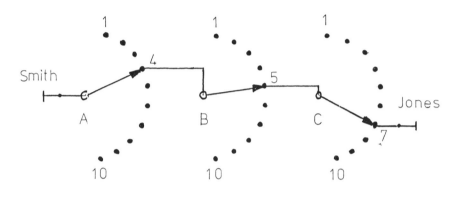

Fig. 5.1   Simple switching system

   Consider the simple system shown in Figure 5.1.   Here
it is assumed there is a switching system comprising three
rotary switches in cascade, each having 10 outlets.   The
object is to connect Mr. Smith with Mr. Jones via switches
labelled A,B & C. Obviously Mr. Jones' number is 457 so
switch A moves to 4, switch B to 5 and switch C to 7.   This
system is clearly a nonsense for it is implied that if Mr
Smith is not making a call the switches A,B & C are idle.
Even worse, the implied suggestion is that Mr. Smith has one
A switch to achieve his choice of hundreds, 10 B switches to
choose tens within each hundred and 100 C switches to choose
the final digit making 111 switches in all.   This is quite
ludicrous - so our user must share switches.

   Hence, let it be supposed that each user is given an A
switch to himself, but say 50 users are grouped together so
that all the contacts with the same number on their A
switches are commoned. We then need just 10 B switches to
serve them all. See figure 5.2.   This is not a very good idea
either as, if any one of the 50 users dials 457 then none of
the remaining 49 can dial any number between 400 and 499 even
though 49 A switches and 9 B switches are idle.

   A much better idea is to put in an extra switching
stage.   Let each user have his own A switch but its use will
be only to find free equipment.   Above, it was assumed the A
switch would respond to the dialled (keyed) digits, but now

the B switch is the first to do so. Thus the user has no
control over his A switch except when he picks up his
receiver, the doing of which initiates the A switch to perform its sole task of finding some free equipment.

Now, as before, it is assumed that the contacts on the
A switches are commoned. When a user picks up his receiver
his A switch will immediately go to contact one. Let it be
supposed that this contact is occupied by another user in the
group of 50 users. Then the A switch will go to contact two
and then three and so on until a free outlet is found. Dial
tone is then fed back and the user proceeds to dial the required number.

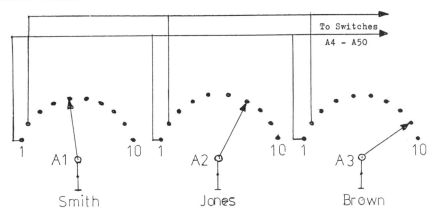

Fig 5.2 Commoning of outlets of a group of switches

Suppose, a minute later, another of the group of 50 users
picks up his receiver. His A switch will go to contact one
which, may by then, be free even though contacts two, three
and four are still occupied by others. This call will,
therefore, be made on contact one as there is no need for the
switch to go any further.

Consider when ten of the group of fifty are making calls
simultaneously, and an eleventh member picks up his handset.
This member's A switch will step round testing each outlet
for availability until the last is reached and, as that is
also busy, a signal will be fed back indicating that there is
no free equipment. At that moment, it could be that one of
the earlier choices has just become free but the switch, as
described, will not step backward and the call attempt has
failed.

With modern delay systems the call failed attempt will be
stored and dealt with when a free outlet is available, and
thus, will not have failed from the user's point of view. In
the simple system being dealt with here, as with the majority
of the public network in use at present, the call is consid-

ered lost. Of course, the user may himself simulate a delay
system by trying again immediately, and getting through, but
in this discussion, that is regarded as a fresh call.

Traffic theory addresses the question, on a probability
basis, "how many eleventh users can be tolerated?"   If
there are too many, as may be the case if all 50 users
require service often, such as perhaps in a sales office, it
may well be that the group of 50 is too large and it should
be reduced to say 30 or 35.   Then, very occasionally, a
failed call will arise but it may be rare enough to be
acceptable.   Conversely, it could be that in a group of 50
users   nobody ever fails to get through, and the last three
or four outlets are never used.   In that case there is an
amount of very expensive equipment always idle and not
justifying its installation as it earns no revenue.   Hence
it may be reasonable to increase the group of users from 50
to say 80 or 90.

It is clear, therefore, that the important factor in
exchange design is not the number of users connected to an
exchange but the traffic they generate.

5.5 DEFINITIONS

A few definitions are essential.

1.   Busy Hour·

This has been dealt with above.   It is simply the hour of
the day when traffic is at its greatest.

2.   Traffic Units

The   unit is named after Agner Krarup Erlang (Danish
1878-1929) and given the symbol E.   Let C calls of average
duration t occur during a period T, then we have

$$\frac{C\,t}{T} \quad \text{Erlangs of traffic}$$

Usually t and T are
measured in hours.

This can be interpreted to mean the average number of calls
in progress simultaneously.

3.   Grade of Service   (G of S)

This is not a measure of the quality of a connection but a
measure of the probability of NOT getting a connection at
all.   It is nothing to do with faults, except in the sense
that a faulty piece of equipment may not be available, but it
simply means the probability of NOT making a connection given
that the incidence of traffic conforms to the statistical
assumptions made, and the amount of equipment available. Perhaps the term 'Grade of Diservice' would better describe it.

Clearly, if the Grade of Service at the busy hour is considered adequate then, during the remainder of the day, the Grade of Service will be better i.e. smaller

5.6 ERLANG'S LOST CALL FORMULA

Before quoting the formula (it will not be proved here) it is important to know the assumptions implicit in it.

1.   Calls that are lost have zero holding time

2.   Pure chance traffic is as likely to arrive at one time as any other time. An infinitely large exchange is implied but a reasonably large exchange is an adequate representation.

3.   The system is assumed to be in statistical equilibrium. This means that while individual calls will begin and end, the overall position of the distribution of calls being processed remains a constant. Naturally, this is assumed to be substantially true over the busy hour.

$$B = \frac{a}{A} = \frac{\dfrac{A^N}{N!}}{1 + A + \dfrac{A^2}{2!} + \dfrac{A^3}{3!} + \dfrac{A^4}{4!} \dots \dots \dfrac{A^N}{N!}}$$

Where N = Number of trunks (outlets or circuits)
                 A = Average traffic
                 a = Traffic lost
                 B = Grade of Service, i.e. the proportion of traffic lost. Equally, it is the probability that N circuits are busy.

Erlang's work was not accepted by his engineering colleagues in the Copenhagen Telephone Company for a long time. In fact he was so little accepted, as a mathematician, that one reads he was not even allowed to join the engineers for lunch!!.

The question he attacked was "Given a route of 20 circuits in which, on average, 12 are busy, what is the probability that all 20 are busy?".

A very simple example may be helpful. Suppose there is a full availability group of 10 trunks which is offered 4 Erlangs of traffic. What traffic is carried by each of the first three trunks and what is the Grade of Service given by the full ten trunks?

NOTE: Full availability simply means that all ten trunks are available to the offered traffic.   This would not be the case if some of the trunks are shared by another part of the switching system.   Using the formula gives:

$$a = \frac{\dfrac{A^{N+1}}{N!}}{1 + A + \dfrac{A^2}{2!} + \dfrac{A^3}{3!} + \dfrac{A^4}{4!} \dots\dots\dots \dfrac{A^N}{N!}}$$

Consider the first trunk, i.e. Put N = 1

$$a = \frac{\dfrac{A^{1+1}}{1!}}{1 + A} = \frac{A^2}{1 + A} = \frac{4^2}{1 + 4} = 3.2 \text{ Erlangs}$$

This means that the first trunk has been offered 4 Erlangs and 'lost' 3.2 of them.   'Lost' means passed on to the next outlet.   Thus the first trunk has carried:-

    4 - 3.2 = 0.8 Erlangs.

Next is calculated what the first two trunks lose to the third i.e. putting N = 2

$$a = \frac{\dfrac{A^{2+1}}{2!}}{1 + A + \dfrac{A^2}{2!}} = \frac{\dfrac{A^3}{2}}{1 + A + \dfrac{A^2}{2}} = \frac{\dfrac{4^3}{4}}{2 \left(1 + 4 + \dfrac{4^2}{2} \right)}$$

    = 2.46 Erlangs

Thus if the pair of trunks have lost 2.46 Erlangs, the traffic carried by the second is clearly 3.2 - 2.46 = 0.74 Erlangs.

By the same process putting N = 3 shows that the third trunk carries 0.66 Erlangs.

Substituting in the initial Grade of Service expression provides a value for B (G of S) of approximately 0.005.   That is to say that one call in 200 would be lost because of insufficient equipment.

It will be noticed that later trunks are less heavily used than early choices. The last trunks in a group will carry very little traffic indeed, and none outside the Busy Hour. However, such trunks cost no less to provide merely because they are less used. Accordingly, it is desirable to load each trunk as fully as possible. Modern systems can use various strategies to achieve a more uniform loading, as they are not restricted to the sequential process used by Strowger step-by-step systems. However, with traditional systems using sequential loading, efficiency may be improved by putting trunks in as large a group as possible. Fig. 5.3. illustrates the effect of this. Efficient loading of trunks, although desirable, is not without disadvantage when overload is considered.

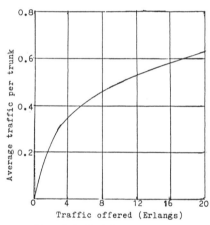

Fig. 5.3 Traffic carried per trunk for a fixed grade of service

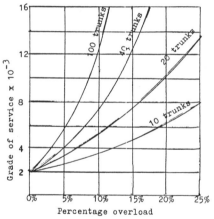

Fig. 5.4 Overload effect on grade of service

A relatively inefficient small group has clearly a low occupancy, and thus an overload will cause a quite small deterioration in the grade of service simply because there is more 'waste' capacity to cope with any unexpected peak in demand. For example a 10% overload does not quite double the Grade of Service for a small group of 10 trunks whereas a similar overload for a group of 100 trunks increases it by over five times.

Hence there are two bases for determining the number of circuits required. First the number to provide the chosen

Grade of Service for a normal busy hour and, secondly, the number required to give an acceptable, but higher, loss when there is a specified overload such as 10%. The appended table gives the number of circuits required to carry various amounts of traffic at Grades of Service of 0.005, 0.002, and 0.001. The standard figure is 0.002 with the additional condition that a 10% overload should not cause the proportion of lost calls to exceed 0.01, that is a worsening by five times. As each criteria will suggest a different number of circuits are required, clearly the higher one must be used to satisfy both the requirements set. See figure 5.4.

Modern systems are more efficient from the circuit utilisation point of view than older ones, but for that very reason they have less 'spare' capacity to cope with overload than some systems they replace.

5.7 THE INPUT PROCESS

Calls arriving at a switching system do so in various ways. They may arrive in a purely random fashion and this is the most common assumption. This may not, of course, always be so, for there may be factors which determine call arrival which are outside the system being considered. For example an overflow system will receive calls only when the first choice system is fully loaded. It may be that in some cases calls will arrive, for some reason, at regular intervals. This is a matter well beyond the scope of this introductory survey. Only one probability distribution which describes these input processes will be briefly looked at. That is the Poisson Distribution.

Assumptions made are as follows:-

1.   The probability that a specified number of calls arrive during an interval of specified duration is the same for all intervals, and is independent of the arrival, or non-arrival, of calls outside this interval.

2.   If the chosen interval is sufficiently small, the probability that it contains more than one call is negligible. This criteria is met by making the intervals as small as is necessary. It is arguable that two calls arriving at the 'same' time, do not in fact do so and that one arrives before the other by some interval, no matter how small. This is illustrated in figure 5.5.

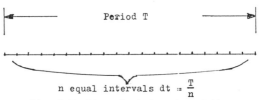

n equal intervals dt $= \dfrac{T}{n}$

Fig. 5.5 Discrete intervals of time

Each interval dt of the period T may, or may not, contain an event (i.e. a call may, or may not, occur during any interval). Each interval is independent. This can be thought of as a series of die throwing to achieve a 6 (call arrival) or not a 6 (no call arrival).

Let p = probability that a particular interval contains a call arrival, then np is the average number of arrivals during period T. This is also AT where A is the average number of calls per unit.

Hence $np = AT$ or $p = \dfrac{AT}{n}$

From the assumptions it can be deduced that the probability of x out of n intervals containing arrivals is:

$$P_T(x) = \underset{n \; \infty}{Lt} \; {}^n C_x \; p^x \; (1 - p)^{n - x}$$

This ultimately evolves to:

$$P_T(x) = \frac{(AT)^x}{x!} \, e^{-AT}$$

An example will be helpful:

A traffic load of one Erlang is offered to a full availability group of three trunks. The average call duration is two minutes.

a.   What is the average number of calls offered per hour?
b.   What is the probability that no calls are offered during a specified period of 2 minutes?
c.   What is the proportion of lost traffic?
d.   If the trunks are always tested in the same order, how much traffic is carried by each trunk?

Solution:

a.   Average number of calls offered per unit holding time = traffic offered in Erlangs (by definition) = 1.
Hence average number of calls per hour = $1 \times \dfrac{60}{2} = 30$

b.   Average number of calls offered per two minutes =1.

$$P_T(x) = \frac{(AT)^x}{x!} \, e^{-AT}$$

But unit holding time = 2 minutes = 1 unit and A = one call per unit time, thus

$$P_T(x) = \frac{(1.1)^{\emptyset}}{\emptyset!} e^{-1.1} = \frac{1}{1} e^{-1} = \emptyset.368$$

c.   Using Erlang , the proportion of lost calls is:

$$\frac{\dfrac{1^3}{3!}}{1 + \dfrac{1^1}{1!} + \dfrac{1^2}{2!} + \dfrac{1^3}{3!}} = \frac{1}{16}$$

d.   Traffic offered to first choice = A_1 = 1 Erlang

Thus probability that 1st is engaged is $\dfrac{1}{1 + A_1} = \dfrac{1}{2}$

Therefore traffic offered to 2nd choice = $A_2 = A_1 \times \dfrac{1}{2} = \dfrac{1}{2}$ E

Similarly putting N = 2 in formula gives probability that first two choices are engaged as 1/5.   Therefore traffic offered to the third choice is

$$A_3 = A_1 \times \frac{1}{5} = \frac{1}{5}$$

Probability that all three choices are engaged we already know from above = 1/16.

Hence overflow traffic = $A_F = A_1 \times \dfrac{1}{16} = \dfrac{1}{16}$

Thus:-

Traffic carried by first choice = $A_1 - A_2$ = $\emptyset.5$ E

Traffic carried by second choice = $A_2 - A_3$ = 0.3 E

Traffic carried by third choice = $A_3 - A_F$ = $\emptyset.138$ E

    This introductory survey has concentrated on loss systems because they are easier to deal with.   Delay systems are not concerned with losing calls but rather with the expected time a user will have to wait for service.   An extension of Erlang's work deals with this matter but such is beyond the level of treatment given here.

Modern systems tend to use non-blocking networks and the limitation is not so much outlets and cross points, but rather the limitation of processors, and modern traffic theory addresses that problem. Those who want to know more are advised to look at the book by Donald Bear on the subject, which is published by Peter Peregrinus, and to whom the writer of these notes is much indebted for his own introduction to the subject.

REFERENCES.

1. Bear, Donald, "Principles of Telecommunications-Traffic Engineering" Peter Peregrinus, 1976.

OUTLINE TRAFFIC TABLE (Full Availability)

No. of Trunks	Capacity for a G.of S. of:-			No of Trunks	Capacity for a G.of S. of:-		
	.005	.002	.001		.005	.002	.001
1	0.005	0.002	0.001	26	15.8	14.5	13.7
2	0.105	0.065	0.046	27	16.6	15.3	14.4
3	0.35	0.25	0.19	28	17.4	16.1	15.2
4	0.70	0.53	0.43	29	18.2	16.9	15.9
5	1.13	0.90	0.76	30	19.0	17.7	16.7
6	1.62	1.32	1.14	31	19.8	18.4	17.4
7	2.16	1.80	1.58	32	20.6	19.2	18.2
8	2.73	2.31	2.05	33	21.4	20.0	18.9
9	3.33	2.85	2.56	34	22.3	20.8	19.7
10	3.96	3.43	3.09	35	23.1	21.6	20.5
11	4.61	4.02	3.65	40	27.3	25.7	24.5
12	5.28	4.63	4.23	45	31.6	29.7	28.5
13	5.97	5.27	4.83	50	35.9	33.9	32.5
14	6.63	5.92	5.44	55	40.3	38.1	36.7
15	7.38	6.58	6.08	60	44.7	42.4	40.8
16	8.10	7.26	6.72	65	49.2	46.7	45.0
17	8.84	7.95	7.38	70	53.7	51.0	49.2
18	9.58	8.64	8.04	75	58.2	55.2	53.5
19	10.34	9.35	8.72	80	62.7	59.4	57.8
20	11.10	10.07	9.41	85	67.2	63.7	62.1
21	11.87	10.80	10.11	90	71.8	67.9	66.5
22	12.64	11.53	10.81	95	76.3	72.1	70.9
23	13.42	12.27	11.52	100	80.9	76.4	75.3
24	14.21	13.01	12.24				
25	15.0	13.76	13.0				

Chapter 6

Customer terminals

Gerry Lawrence

6.1 INTRODUCTION

The term customer terminal is used to describe the apparatus which is connected to the local network and which resides at the customer's premises. The terminal may be connected to the local network by means of a PABX, or other customer switching system, but this chapter does not concern itself with those terminals which are specific to one particular type of PABX or key system. By far the greatest number of any one type of terminal is the telephone, connected to the Public Switched Telephone Network (PSTN). Details of the telephone are given, together with those of other terminals on different networks, for example Telex terminals. The services which are currently provided by several different networks are coming together into one network, known as the ISDN - Integrated Services Digital Network (for details, see Chapter 12). This chapter concludes with information relating to terminals for use on the ISDN.

6.2 TELEPHONES

The word telephone is derived from the Greek words 'TELE - at a distance' and 'PHONE - sound' and is used to describe an instrument which enables people to talk over long distances. The use of telegraphy or Telex permits the exchange of messages, but when carried out in real time the information rate between the two people communicating is very much lower than with speech. There are some important features of speech communication missing from telegraphy, such as the ability to impart emotion into the message by the use of inflection of the voice. Also it is possible to confirm the identity of a speaker over the telephone by the characteristics of the voice, even though the speech transducers and transmission medium employed impart considerable distortion to the original signal.
The telephone was invented in 1876 by Alexander Graham Bell, and its first realisation was an electrical equivalent of the 'cocoa tin and string' method of conveying speech. A pair of Bell electro-magnetic receivers were connected together over two wires, each transducer being used

alternately as a transmitter and a receiver. The sound pressure waves from the person speaking moved a diaphragm which by means of the electromotive force principle caused a varying electric current to flow along the wires. This was applied to the receiver which used the inverse electromotive force principle to cause a diaphragm to generate sound pressure waves for the listener. Distances of up to a few miles could be achieved with this technique and the fundamental requirement of telephony had been realised.

Apart from the fact that speech was only single point-to-point, there were a number of operational drawbacks with this system. It was not possible to talk and listen at the same time, but to a large extent this was overcome by the use of verbal handover procedures. A much more serious deficiency existed in that it was extremely difficult to actually initiate a telephone conversation without the prearranged co-operation of the other person. Since the sound level emanating from the receiver was not loud enough to attract attention, a call alerting device was required. One method employed was to leave the receiver with the diaphragm facing upwards with a ball bearing resting on it. This could be made to rattle by means of an acoustic signal generated at the other end. An alternative arrangement was to fit a bell at each end. This was activated by the call originator who generated the appropriate AC voltage to drive the bell by turning the handle on a magneto generator. Thus it was established that the minimum functional requirements of a workable telephone instrument are for four discretely identifiable communication channels :-

    (i)   transmit speech
   (ii)   receive speech
 (iii)   outgoing signalling
  (iv)   incoming signalling

6.2.1 Transmit Speech

The early Bell electromagnetic transmitters were not very sensitive, that is they did not generate a very large signal for a given sound pressure level. In 1877 Thomas Edison invented the carbon transmitter which had an inherently higher sensitivity. Not being an EMF generator it required a separate DC power source feeding current through the transmitter. The movement of the diaphragm altered the resistance of the transmitter, thereby superimposing on the DC feed an AC signal; this was inductively coupled into the main circuit by means of a transformer. The DC power source was obtained by means of a local battery (LB) fitted to the telephone; subsequent development led to the provision of a Central (or Common) Battery in the exchange (CB).

The carbon transmitter has remained for many years the most commonly used form of microphone, and there are many millions in service throughout the world. However, it suffers from a number of disadvantages. The resistivity and therefore the sensitivity is very dependent on the packing

together of the carbon granules. During manufacture this is
very closely controlled but in service an ageing process
occurs, which is accelerated by use, temperature and
humidity resulting in a gradual decrease in sensitivity.
It is well known that the sensitivity can be temporarily
increased by giving the handset a sharp jolt. Another
result of the ageing process is an increase in the amount of
noise present in the form of hissing and crackling. In
consequence, the carbon microphone is rarely employed in new
designs of telephone, where it is more common to find a
dynamic microphone. This uses a transducer which generates
a small electrical signal that can be amplified to give the
correct send loudness rating. Transducer drive units
currently make use of one of the following principles :-

* charged electret foil (condenser microphone)
* moving coil
* moving iron or moving magnet
* piezo - ceramic foil
* piezo - ceramic disc.

Amplifier circuits draw their operating power from the
line, and are generally multiple transistor stages giving an
overall voltage gain of about 23dB. The amplifier is often
incorporated into an integrated circuit. Because of the
cost of this amplifier, electronic transmitters are more
expensive than carbon microphones, when used as a direct
replacement. Nevertheless, they have a better service
record in relation to overall reliability and the meeting of
performance specifications throughout the life of the
transmitter. Where other functions within the telephone are
provided by electronics, the microphone amplifier can be
included at little or no cost; under these circumstances the
electronic microphone is cheaper than the carbon equivalent.
It does, however, suffer from a number of disadvantages :-

* care must be taken to ensure that the amplifier does
  not pick up RF signals, for example from radio
  stations.

* the microphones generally have a linear sensitivity
  with decreasing sound pressure input, and this leads
  to a greater level of ambient room noise transmitted
  to line than for carbon transmitters.

* attention must be paid to the circuitry to ensure
  that two telephones can perform satisfactorily when
  connected in parallel to the same line.

6.2.2   Receive Speech

Electromagnetic principles for converting electrical
energy to sound energy have dominated the design of the
receiver since the invention of the telephone, and the
majority of development has been directed towards improving
the frequency response and increasing the sensitivity.

Various techniques have been employed including moving coil
and moving iron and it is a version of the latter, the
rocking-armature receiver, which is in use today in the
majority of the world's standard telephones. The receiver
contains two windings which in response to the AC speech
signal pull and push an armature that is free to rock on a
knife edge bearing. The armature activates a metal
diaphragm to give the desired sound output, and the degree
of magnetisation of a permanent magnet sets the sensitivity.
    Modern telephones employ alternative forms of receiver
which make use of an amplifier in the receive path to enable
a transducer to be used with an inherently lower
sensitivity. This in turn leads to wider tolerances on
parts and cheaper assemblies. Examples are :-

- lower sensitivity rocking armature
- moving coil
- moving iron disc armature
- piezo-ceramic disc

6.2.3   Speech Circuits

    Shortly after the invention of the carbon transmitter
the disadvantage of half-duplex speech in Bell's original
telephone was overcome by fitting each instrument with both
a transmitter and a receiver. Although simultaneous talking
and listening was now possible a new problem had been
introduced, namely sidetone. This is the name given to the
basically undesirable sounds that are reproduced in the
user's own receiver, and which did not originate from the
far end. For example, a listener will now not only hear the
far end speech but his receiver will reproduce any room
noise picked up by his own transmitter. This is known as
'listener sidetone' and it can be reduced during listening
by covering up the mouthpiece. On the other hand, a talker
will now hear his own voice reproduced in his receiver and
this is known as 'talker sidetone'. This is in addition to
the natural sidetone which comes through the air from mouth
to ear and through the bones in the head. Talker sidetone
can be reduced by holding the earpiece away from the ear,
although this is less easily achieved with one piece
handsets than was possible with telephones of the
candle-stick type.
    Sidetone in early telephones was not a major problem
because of the low sensitivity of the receiver. However,
the problem increased as the sensitivity increased. The
human actions that were taken to reduce the sidetone started
to detract from the benefit of having both a transmitter and
a receiver and so a solution was sought to perform the
anti-sidetone function electrically. The first circuit is
attributable to Campbell and Foster in 1920 and is based on
the principle of using a triple-wound transformer to cancel
out mutual effects between transmitter and receiver. The
hybrid transformer is designed to carry out two-to-four wire
conversion. That is, it couples the two wires of the line
(carrying speech in both directions) to the pair for the

transmitter and the pair for the receiver, segregating the speech signals as it does so.   It is, however, complicated by the following factors :-

• DC current from the line must be fed to the carbon transmitter, but not to the receiver.

• Not all of the talker sidetone should be cancelled, since the user will complain of a 'dead' telephone. (Sidetone is discussed in Chapter 2).

• The level of sidetone achieved is determined by the success of an RC network in balancing the impedance of the telephone against that of the line.   Since the line impedance varies with length, and both impedances vary with frequency the balancing network is always a compromise.

The result is the classical telephone circuit shown in Figure 6.1.

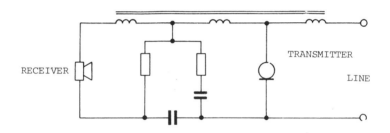

RECEIVER

TRANSMITTER

LINE

Fig. 6.1   Classical Speech Circuit

Another function of the speech circuit is to balance the send and receive line sensitivities so that the correct overall send and receive loudness ratings from mouth to ear are achieved (see Chapter 2).   Normally, because of the high sensitivity achieved by the rocking armature receiver, the speech circuit has little effect on the transmit level sent to line at the expense of a loss in the receive signal presented to the receiver.

The third requirement of the speech circuit is to provide the function of regulation.   Having striven to raise the send and receive sensitivities to cope with ever increasing local line lengths, it is then found that the performance of the telephone is far from ideal on a very short line.   Transmit and receive levels are too loud and sidetone levels are too high.   One way of overcoming these difficulties is to introduce a circuit whose impedance decreases as the DC line current increases and use this impedance to shunt the transmitter; a similar circuit is also needed for the receiver, although the loss introduced is less than that for the transmitter.   The regulator circuit must cater for both polarities of line current, and is normally designed to be active only below about 3km. The combination of anti-sidetone and regulation leads to the standard speech circuit as shown in Fig. 6.2 in use in present 700 series and other non-electronic telephones.

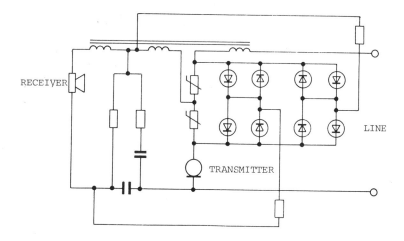

Fig. 6.2 700 Series telephone speech circuit

Over the last decade there has been an increasing use made of electronics in telephones particularly in the form of LSI integrated circuits. These circuits fall into three categories :-

(i)   circuits to carry out basic analogue speech functions

(ii)   circuits to carry out basic signalling functions

(iii) circuits to provide in the telephone additional features, such as memory or display

Transformers can not easily be incorporated into integrated circuits and this has led to the hybrid transformer being replaced by a bridge network consisting of resistance and capacitance (commonly known as a resistive hybrid). Amplifiers are associated with both the transmitter and the receiver (leading to lower inherent sensitivity in each transducer) and are connected across the two opposite points of the bridge. The amplifiers supply gain to both send and receive speech signals to achieve the correct loudness ratings, and have two degrees of gain control applied to them :-

(i)   derived from the loop current in order to apply regulation according to the length of the line.

(ii)   derived from the speech waveform to limit the
maximum signal sent to line (transmitter), and to
prevent acoustic shock (receiver).

Sidetone is controlled by the balance of the bridge at
each frequency and line length, and in any practical
realisation of the circuit this must inevitably be a
compromise, as with the hybrid transformer approach.
The impedance of the telephone presented to the line is
important, in order to achieve a good match to the local
network characteristics. The 700 series with its hybrid
transformer presents a complex impedance that is not a very
good match to the local cable, and which varies with loop
current and frequency. With electronic telephones employing
resistive hybrids it is much easier to design a complex
impedance speech circuit that is more constant, and thereby
achieve a better sidetone balance over a wider range of line
lengths and signal frequencies.

6.2.4 Outgoing Signalling

The need has already been established for a user to
indicate that he wishes to initiate a call. In Bell's
original point-to-point set up, a magneto generator alerted
the other end by means of a bell. It rapidly became
appreciated that point-to-point telephone connections were
of only limited appeal, and soon operator controlled
switching centres (exchanges or central offices) were
introduced to enable calls to be set up to any subscriber
connected to that exchange. Initially the magneto generator
was used for alerting the operator, and she in turn was
responsible for alerting the called user. With the
introduction of CB (Central Battery) exchanges, the alerting
signal to the operator became a simple loop across the line
at the telephone; the flow of DC current was used to trip an
indicator on the operator's control panel. Thus the
generator could be eliminated from the telephone, since only
the operator now needed to send ringing current along the
lines.
When a Kansas City undertaker named Strowger invented
the automatic exchange in 1891 (in order, so it is said, to
be able to dispense with an operator who was directing
business meant for him to one of his rivals) it became
necessary to send a sequence of digits to the exchange to
indicate the called party to whom he wished to be connected.
This was done by interrupting the loop current a number of
times in succession equal to the value of the digit required
(0 being equal to 10 pulses). The current must not be
interrupted for too long, or the exchange will think that
the subscriber has cleared down; and the make period between
breaks must not be too long, or the exchange will think the
subscriber is pausing between digits. These two
requirements led to a signalling speed of 10 impulses per
second, taking one second to signal digit 0, and known as
10 i.p.s. or Decadic Signalling.
A rotary dial was chosen as the most appropriate

mechanism. Because on long lines the break pulses tend to get shorter when detected at the exchange compared with those originated at the dial, it is necessary to apply pre-distortion; in place of a 50/50 break/make ratio the dial generates pulses with a duration of $66^2/3$ ms break and $33^1/3$ ms make (60/40 in the USA). The dial contains a clockwork mechanism, a slipping clutch during wind-up and a governor to control the return speed to within 9-11 impulses per second. In addition to breaking the loop current it has off normal make contacts for the following functions :-

- to provide a short circuit path through the telephone to help prevent impulse distortion.

- to shunt the receiver, to prevent unwanted loud clicks in the earpiece.

- to shunt the transmitter to avoid pulsing DC through the carbon granules and thus cause premature ageing.

The rotary dial has now been replaced by a push button electronic circuit which performs exactly the same functions. An integrated circuit stores the digits keyed in, and controls all the timing functions necessary to achieve the correct impulsing of the digit sequence with the appropriate pauses inserted. The pulsing and off-normal functions are carried out using transistors as shown in the schematic circuit in Figure 6.3, which is the signalling circuit employing a custom-designed CMOS chip. One design problem is that the chip must continue to function when the loop is alternately open and short circuit and for this purpose a storage capacitor (Cl) is provided.

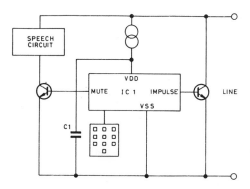

Fig. 6.3   Push button 10 i.p.s. signalling circuit

Early push button 10 i.p.s. telephones relied on rechargeable batteries for this power which added to the cost and complexity, and therefore these telephones did not achieve popularity. Apart from satisfying a demand for a push button instrument, this type of telephone appears to offer no benefits to the user; on the contrary he has to suffer a very long post-dial delay before the network responds. However, its advantages are :-

- it can be fitted to any existing exchange line without modification to the exchange.

- the impulses sent to the network are more accurately controlled than from a dial, leading to fewer wrong numbers of the sort caused by the network misinterpreting the digit sequence.

- in certain circumstances there may be a small reduction in the overall dialling time due to the reduced 'thinking time' needed by the subscriber to input all the digits into the telephone.

- by the use of low power CMOS technology it is possible to incorporate into the integrated circuit features such as Last Number Re-dial and Repertory dialling.

In order to speed up the process of transferring digits from the telephone to the exchange an alternative to 10 i.p.s. signalling is required. In some networks loop disconnect signalling at 20 i.p.s. has been used, but cable problems can be encountered at these speeds. An alternative form of signalling makes use of the audio band to send multi-frequency (MF) tones from the telephone to the exchange. Each digit signal consists of two tones, selected one from a low band and one from a high band set of frequencies. These are shown in the table below, and this form of signalling is referred to as MF4.

		HIGH GROUP (Hz)			
		1209	1336	1477	1633
	697	1	2	3	A
Low Group (Hz)	770	4	5	6	B
	852	7	8	9	C
	941	*	0	#	D

The digits A, B, C, D are not in common use at this point in time in public telephone networks. Two tones not related harmonically are employed for the signalling in order to reduce the possibility of imitation of digits by the human voice. A simple circuit in the telephone consisting of a single transistor oscillator coupled to various inductance and capacitance combinations has been used since the early 60's. This circuit has two disadvantages :-

(i) the frequencies generated are not completely stable with respect to temperature and so the tolerance of each individual frequency allowed by the telephone exchange receiver can under certain circumstances be exceeded.

(ii) the timing of the two tones is largely dependent on the mechanical action of the push button mechanism for minimum duration. Maximum duration is usually determined by how long the button is held down by the user.

Both of these disadvantages may be overcome by the fitting of additional electronic circuitry. A better approach is to employ an integrated circuit to carry out the tone generation. The I.C. is normally provided with a minimum of 4 digits storage and can thus cater for a fast rate of digit input while maintaining output tones with a minimum duration of 70 ms. The maximum duration can also be timed by the integrated circuit although some administrations prefer to use the traditional untimed method. Where the integrated circuit provides for last number re-dial or repertory features, the durations of the tone burst and the silent period are both controlled by the chip. It is normal to provide for a unity mark/space ratio, with each period timed to last between 70 and 100 ms.

Although it is necessary to mute the microphone during tone sending, to avoid picking up sounds which might confuse the MF receiver at the exchange, most telephones provide for a low level feedback of tones to the receiver to act as a 'confidence tone' to the user.

Although the MF telephone overcomes the limitations of the slow speed signalling of the 10 i.p.s. push button instrument, it does suffer from the disadvantage of requiring specialised exchange equipment capable of receiving, filtering and de-coding the pairs of tones signalled to it. Not all exchanges are fitted with this equipment and therefore to permit the use of MF telephones it is necessary in these cases to add apparatus which carries out the function of tone to pulse conversion. Whether or not there is widespread application of this conversion equipment depends on a number of factors, including how many exchanges there are in the network which do not already have an MF capability. The current policy in the UK is to fit tone to pulse converters on a relatively limited scale and rely on decadic telephones to satisfy the demand for push button instruments. All modern PABX's cater for MF signalling, and System X exchanges currently being installed in the PSTN rely on the provision of an MF capability in order to be able to offer the user a range of specialised features known as 'star services'. Because MF4 signalling is totally in-band it may be used as an end-to-end slow speed signalling system. This has found applications in banking, for example, where after dialling through to a bank, the user may input various account and security code numbers by means of the telephone keypad. It is usual in these cases for the response to be in the form of synthetic speech generated by a computer. An extension of this idea is used in voice messaging systems. The MF4 signalling is first used to control the deposit of a voice message into a memory; the message is generated by the caller talking in the usual way. Subsequently, the message may be replayed to another caller if he is able to supply the correct password using the MF4 keypad.

These types of features are not directly available to the vast majority of users who are connected to exchanges which only accept 10 i.p.s. signalling. However, for such people telephones are available which can convert at the touch of a button from decadic to MF; it is common to use the * or # button to achieve this, since they are redundant in decadic signalling. These telephones are also useful in the situation whereby such a user wishes to access the Mercury network from the BT network. The latter may require 10 i.p.s. signalling, but Mercury requires the identity of the calling and called user to be input in MF form.

Once the digits have been transmitted from the telephone to the exchange, the subscriber then enters the speaking mode and this normally persists until the end of the call. However, there are cases when it becomes necessary in the middle of the speech phase to recall the central register equipment, for example when wishing to bring a third party into a conference. For this purpose the telephone is fitted with a recall button which provides one of two different signals :-

(i) one of the two wires connecting the telephone to the exchange is short circuited to earth through a contact. This method is common on PABX's and is known as earth recall.

(ii) a timed break recall pulse consisting of a disconnect signal (the same as a dial break) is transmitted to the exchange. For decadic telephones this may be obtained by simulating the operation of button 1 but this method is not suitable for MF telephones. In the latter case a circuit has to be fitted which provides a break pulse of controlled duration when the recall button is activated. Sometimes the duration required is greater than 150 ms, but this must be closely controlled in order not to imitate the normal clear down signal presented to the exchange. Timed break recall is usually fitted to public exchanges in preference to earth recall to avoid the problems of earth potential difference between the exchange and the telephone when the user is at some distance from the exchange. However, some PABX's offer recall in the form of a 'switch-hook flash'. This is a relatively loosely toleranced timed break, and only works in conjunction with a long cleardown time-out.

When the subscriber wishes to terminate a call he normally replaces the handset which sends a continuous break to the exchange. The exchange will assume the call has been abandoned and that a clear signal has been sent when the loop is broken for more than 180 ms. PABX's offering switch-hook flash facilities can require a cleardown signal of half a second.

The only remaining outgoing signal not mentioned so far is that which is sent to the exchange in order to answer an incoming call. This of course is the same as the call initiating signal, that is a continuous loop presented to the exchange. The exchange is designed to detect the DC path established through the telephone while it is sending the ringing signal.

6.2.5 Incoming Signalling

The telephone is required to receive an incoming signal which indicates that a call is arriving and to alert the subscriber to this fact. The signal presented to the telephone is incoming ringing current and in early manual exchanges used to be derived from a magneto generator. With the advent of automatic telephone exchanges this signal had to be applied without human intervention, and for this purpose a ringing machine provides the basic signal to which automatic connections are made at the appropriate time. The signal generated consists of 75v rms at either $16^2/3$ or 25Hz in the UK, and is applied to the line to create a ringing cadence that implies urgency. Other countries use different frequencies, such as 20 Hz, and different cadences. PABX's use different ringing cadences to indicate to the user the difference between an internal call and an external call. The ringing current is supplied to the ringer circuit in the telephone through a capacitor whose function is to block the DC path. The hook switch is designed so that in the on-hook state the ringer circuit is connected to the pair of wires and the speech circuit is disconnected. When the user goes off-hook to answer the call the speech and dialling circuits are connected across the pair and the ringer is disconnected; this action provides a suitable low impedance path through the telephone in order to activate a circuit in the exchange known as ring trip. The exchange detects ring trip and removes the ringing current.

Telephones have traditionally used an electro-magnetic device as the transducer to convert ringing current to an acoustic alerting signal. The basic form is that of a coil which activates a hammer to strike a gong thereby translating the subaudible 25 Hz into a signal in the audio band. Part of the urgency of the alerting signal is derived from the fact that the gong creates a strident tone with multiple harmonic content. Alternative forms of call alerting devices use an electronic tone caller circuit connected to a suitable acoustic transducer. These circuits use the power contained within the ringing current but not the frequency. They normally rectify the AC signal and use the derived power to drive an oscillator circuit whose frequency is determined by the components used. The output may be somewhere in the range 1 -1.5kHz consisting of two fundamental frequencies with a suitable low frequency warble (for example 15 Hz) between the two; it can be made to give a tone of similar urgency to that obtained from the normal bell.

When two telephones are connected in parallel to the same telephone line, the bell circuit of a telephone in the on-hook state is across the speech circuit of a telephone in the off-hook state. Therefore the ringer circuit must be designed to appear as a high impedance circuit to speech signals. Electro-magnetic ringers are by their very nature highly inductive and therefore can be made to present a high impedance at frequencies in the audio band, and a low impedance to the 25 Hz ringing current. An alternative method, used particularly with electronic tone caller circuits, is to employ two zener diodes which are able to discriminate on the basis of voltage not frequency. Since speech is at a very low voltage compared with the signal from the ringing generator, the two diodes can be set at a threshold so that the circuit appears to be high impedance to speech and low impedance to incoming ringing.

Telephones in parallel also present a problem known as bell tinkle. Since dial pulses are at a nominal frequency of 10 Hz and are interrupting a DC loop with 50 volts at the exchange, and since ringing current is normally at $16^2/3$ Hz it is possible for the ringer circuit of an on-hook parallel telephone to interpret a long string of dial pulses as incoming ringing; this can cause the bell to tinkle or the tone caller to chirp. It is possible to overcome this in bells by employing a biassed hammer used in conjunction with the fixing of the polarity of the pair throughout the local network, as in the USA. However, BT do not operate this principle in the UK, and alternative solutions have been adopted. If the tone caller circuit employs an integrated circuit to generate the waveforms it can be designed to incorporate a very accurate filter in order to carry out the discrimination. An alternative approach is to use a multiple wire interconnection technique in order to eliminate bell tinkle.

The method currently being adopted in the UK, implemented using the new UK plug and socket arrangement, is to use ringers of approximately 4,000Ω impedance and to put each in parallel using a simple parallel domestic wiring scheme. It is possible to connect up to four bells in parallel, or more if callers of a higher impedance are used. Bell tinkle is eliminated by using a 3-wire parallel interconnection as shown in Figure 6.4. The 1.8μF bell capacitor is fitted in the first terminal block; subsequent parallel sockets do not have components fitted.

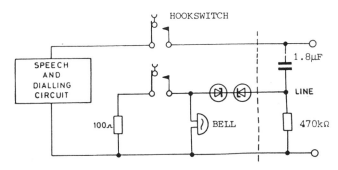

Fig. 6.4   UK 3-wire interconnect circuit

It is not normal for call charging meter pulses to occur in the local network. However, by means of additional equipment in the exchange, it is possible to supply Subscriber Private Meter pulses (SPM) to customer's premises, where they can be used to drive an appropriate meter. In the UK, these take the form of 50 Hz longitudinal signals lasting for 250 ms, and are therefore silent to both calling and called party.

6.2.6   Loudspeaking Telephones

There are two types of telephone within this category :-

    (i)   those in which the received speech is amplified to a normal direct voice level, but the handset is still used for transmit speech.

    (ii)   those telephones which provide for full hands-free operation.

In the first group is a telephone with the facility known as 'on-hook dialling'. The user imitates the action of going off-hook by pressing a seize button which loops the line to the exchange and connects it to an amplified receiver. Forward signalling takes place in the usual way but with the user able to monitor the progress of his call through the loudspeaking transducer. Only when the call has progressed to the stage of two-way speech is it necessary for the handset to be lifted off-hook. Careful design of the amplifier and the seizing circuits enables the feature to be entirely line-powered without exceeding any of the network limitations. It is usually necessary to disconnect the amplifier circuit when the handset is lifted in order to prevent acoustic feedback between the loudspeaker and the transmitter causing a 'howl'. If this is not carried out then in the state just prior to the occurrence of oscillation the loudspeaker will couple received speech into the transmitter, and this might be perceived by the far end as an undesirable form of echo.

This problem is potentially present in the full hands-free loudspeaking telephone, and may occur for two reasons :-

(i) acoustic coupling as above between the microphone and the loudspeaker. This effect may be minimised by careful physical design, since unlike on-hook dialling the spatial relationship between the microphone and the loudspeaker is fixed by the design. However, a resonant room will always provide paths due to reflected sound.

(ii) electrical coupling due to the inefficiency of the two-to-four wire hybrid transformer in cancelling out sidetone.

A common technique to eliminate both of these effects is to use a voice-switched circuit. The telephone operates in the half-duplex mode with the direction of transmission controlled by comparing the level of transmitted speech with that of received speech and switching in favour of the loudest. Hysteresis is built in to avoid rapid switching but this has the undesirable effect of introducing syllable clipping while the circuit is changing its direction of transmission. An alternative method is one based on echo cancelling techniques but since the circuits must be adaptive to cater for dynamically varying conditions this method is currently very complex.

It is common for the full loudspeaking telephone to be mains powered in its full feature mode, but still be able to make and receive calls using a handset in the event of mains failure. However, the use of CMOS technology has enabled designs of LST to be realised that work satisfactorily using only line power, particularly over short to medium line lengths.

6.2.7 Additional Features

Apart from the styling of telephones, for which there appears to be no limit to the artistic ingenuity of the designer, the user has a wide range of facilities from which to choose. The addition of RAM enables the user to store in the telephone his most frequently used numbers, and can also provide 'save' and 'scratchpad' features. Telephones with digit displays enable the user to check the contents of memory locations before the numbers are sent to line, as well as display digits as they are keyed in. Because of power constraints, the liquid crystal display has emerged as the most popular; it is common for the display to give time and date indication when not required for telephony purposes.

Another feature, which is particularly useful for the business user, is the ability to try a number again automatically on encountering busy tone. This is popularly known as 'busy busting'; regulations currently limit the number of repeat attempts that may take place without manual intervention. Some telephones provide the ability to carry out call logging functions at the terminal end. The digits dialled can be recorded locally and the charging information can be determined accurately from the Subscriber Private Meter pulses transmitted from the exchange. However, this type of telephone may become less popular with the advent of itemised billing, which can be provided on System X exchanges.

6.2.8 Payphones

Telephones for use in public call offices, or on private premises where immediate payment for the call is required, are commonly known as payphones. The most widespread form of payphone accepts various denominations of coins, and has evolved through three forms :-

(i)   pre-payment boxes.
(ii)   pay on answer boxes.
(iii)   pay at any time boxes.

Type (i) boxes were very common as the Button A and B boxes up to the end of the fifties, but did not allow for the insertion of additional coins when the paid time had elapsed (except on operator controlled calls). They were not suitable for subscriber dialled trunk calls (STD). As a result, Type (ii) boxes were introduced, which enable a call to be extended by the repeated insertion of coins. They in their turn have been superseded by electronic forms of payphone of the third type in which the amount of credit still outstanding is displayed, and any unused coins are returned at the end of the call. These boxes require a minimum pre-payment coin and accept further coins at any stage in the call.

Payphones require all the speech and signalling facilities of ordinary telephones, but in addition require signals to reconcile the charge for the call with the money inserted. A fundamental decision must be taken by the administration concerning the level of complexity it desires in the coinbox and in the exchange; it is usually the case that the simpler the coinbox the more complicated the exchange equipment is, and vice versa. The pay on answer box is of the simpler type that signals forward to the exchange the value of each coin inserted using 5kΩ loop pulses. This technique requires the association of very complex Coin and Fee Check (C and FC) equipment in the exchange to carry out the call accounting necessary between the value of the coins inserted and the rate of charging of the call. To prevent coins from being inserted before the C and FC is ready, the coin slots at the box are locked until a line reversal is sent to unlock them just prior to the application of pay tone. The slots are locked again when coins to the maximum amount of credit have been registered at the C and FC, and at the end of the last warning of pay tone when the call is to be cleared. This system permits the insertion of coins at any time during the call, but suffers from the disadvantage that the speech transmission must be interrupted while the coin value is signalled to the exchange. On local calls and off-peak trunk calls this is not a major problem, but the situation is becoming intolerable due to inflation and the advent of international subscriber dialled calls.

The alternative arrangement for pay at any time boxes makes provision for silent meter pulse signals to be sent from the exchange to the payphone, with the latter carrying out the call accounting function. The meter pulses may be signals at 50 Hz, or tones of 12 kHz or 16 kHz depending on the telephone network. Coins inserted prior to the call being established, and during the call, are held in suspension until the control logic within the payphone determines that they need to be collected. Coins remaining in suspension are returned to the user when the payphone handset is placed on-hook. When there are no more coins held in credit and the next meter pulse arrives, the payphone supplies pay tone to the user and clears the call if coins are not inserted within the allowable period. It occasionally occurs that only part of the value of a coin held in suspension needs to be collected at the end of a call. If the user places the handset on-hook, then the coin is collected, and the unused value is lost. However, with this type of payphone a 'follow on call' button is provided; this enables the user to contribute the unused value of the coin to an immediately following call, rather than lose it altogether.

Such an arrangement reduces the complexity of the exchange equipment while at the same time making the coinbox more complex; this is less of a problem now that extensive use is made of low power microelectronic circuitry. It has the advantage that coins can be inserted at any time without interfering with the conversation, since coin value signals are not forwarded to the exchange. However, this becomes a disadvantage on operator assisted calls where the operator needs to know that coins to the correct value have been inserted before the call is established. On the modern electronic payphones, the box detects that a call has been made to an operator and modifies its own mode of working. In this state the coins are not held in suspension but drop straight through the mechanism generating audible tones according to the value of the coin. These tones are fed forward to the operator over the normal speech path, and are used for identification and call accounting by the operator. This type of payphone is replacing the present pay on answer boxes in the UK.

As an alternative to accepting coins in payment for a telephone call, some types of payphone are designed to interwork with plastic cards. The most common form in the UK uses a debit card, which can be purchased over the counter, and is initially worth a fixed value. The phonecard is inserted into the reader and the number of call units determined for which the card is valid. As the call progresses, units are erased from the card at each meter pulse. When the call is complete the card is returned to the user, together with an indication on the payphone of the number of units remaining on the card. When all units have been used, the card becomes worthless and may be scrapped; there is currently no way of 're-charging' such cards with more units.

A more convenient method for the user is to employ one of the standard credit cards, but this gives the administration additional problems, particularly if the card is to be used without operator assistance on the call. For example, before the call is made BT must establish the credit worthiness of the user, and perhaps impose a maximum sum for the call. This involves a connection being made to the relevant credit card computer, which also verifies the identity of the user (by means of a PIN input from the keypad). After the telephone call has been made, a second connection with the computer is necessary to debit the account of the cardholder. The payphone may accumulate a number of transactions in its memory before making this connection.

6.3 ANSWERING MACHINES

Many business and domestic users like to offer their callers the opportunity of leaving a message when they themselves are away from the telephone, and this is possible with an answering machine. The basic requirements of this terminal are as follows :-

- the terminal is equipped with a ringing current detector, to recognise the presence of an incoming call. After several cycles of ringing current (to avoid responding to false incoming calls) the answering machine disconnects the detector and loops the line.

- the terminal sends a message to the caller indicating the presence of an answering machine and inviting the caller to leave his own message. The machine's words may be fixed by the design, or may be varied by the owner. In the latter case, storage on magnetic tape or in RAM is provided, and enables the owner to change the message if circumstances require it.

- storage must be provided for the caller's message, and this is usually on magnetic tape. Means is provided to inhibit the answer signal when all the tape has been used.

- the answering machine stops the tape and drops the line if no incoming speech is received over a set period.

A feature which some machines offer is the playback of messages over the PSTN under the control of the owner.

A call is made to the answering machine in the normal way,
and the owner sends an identity signal by means of an
acoustic device held to the microphone of the telephone he
is using. This causes the machine to rewind the tape and
replay any stored messages. The machine reverts to normal
mode of operation when the line is dropped.

6.4 MODEMS

By the use of modulation and demodulation of suitable
carrier tones, whose frequencies lie within the bandwidth of
the PSTN, the telephone network is used to convey binary
information instead of speech. This enables connections to
be made between computers and their peripheral equipment, or
indeed between two computers, with the simplicity of making
a telephone call, and with the benefit of the high level of
penetration of the PSTN into customer premises. In its
basic form, the modem contains a 2/4 wire hybrid, to
separate the two directions of data transmission. The
transmitted data is derived from digital signals in binary
form and used to modulate the carrier in a predefined manner
in order to transmit binary information. For example, in
the simplest form of modem, the carrier is switched between
two different frequencies depending on whether the data is 0
or 1 at any point in time. The level of the signal is
substantially constant, at a value consistent with that of
ordinary speech, so as not to cause problems with cable and
electronic equipment encountered in the PSTN.

At the receiving end, the received signal is filtered
to eliminate any interfering effects caused by the tones of
the return channel, and used to drive a square wave
generator in order to reconstitute the original data signal.
In practice, the received signal will be distorted to some
degree by the PSTN and it is necessary at medium bit rates
(up to 4800 bit/s) to apply equalisation in order to
counteract these effects. For higher speed modems, the
equalisation has to be adaptive. This may require the
sending end to transmit a known data stream while the
receiving end electronically adjusts its equaliser in order
to receive the data pattern correctly. This may take
several hundred milliseconds and is known as the 'learning
period' or 'training period'. Further details concerning
modems are given in Chapter 3.

Using the techniques described, modems with bit rates
up to 9600 bit/s can be connected to the PSTN. They do not
generate data themselves, but are incorporated into many
other terminals which do; some of these are described in the
following paragraphs.

6.4.1 Transaction Terminals

In order to overcome fraud, many shops verify the
validity and authenticity of a credit card at the point of
sale. One method is to make an ordinary telephone call to
the credit card company to establish credit level and obtain
an authorisation code. Once received, the paperwork is

completed in the usual manner.
        An alternative method uses a transaction terminal.
A call is set up directly to the computer and the credit
card is wiped through a reader, sending full details of the
owner. To this is added the cost of the goods and the
user's PIN, and in return an indication is given whether or
not the sale should go ahead. If so, then confirmation
causes the account to be debited directly, and no further
paperwork is necessary.

6.4.2 Data Terminals

        Apart from terminals such as the VDU, modems have been
incorporated into screen based telephone plus data terminals
in order to provide an executive with integrated voice/data
communications. The data section can be connected to a
second telephone line so that simultaneous voice and data
may take place to different locations. For example, the
user may set up a data call to a computer and a voice call
to another executive in order for them to discuss the
information displayed. When the screen is not in use on a
data call, it is often employed to provide the user with
local terminal based features, for example repertory store
or electronic diary.

6.4.3 Facsimile Machines (Fax)

        The PSTN is widely used to copy documents from one
location to another. The document being copied is optically
scanned line by line, and the light and dark areas are
converted into digital information. This is transmitted
over the network using modems and converted at the receiver
into a form suitable for creating a hard copy. Simple Fax
machines operate slowly, taking up to six minutes for one A4
page, and possess few features. The more sophisticated
machines are able to skip over blank lines (at the same time
informing the other end) and reduce the transmission time
for a moderately dense A4 sheet to twenty seconds. They can
automatically adjust their line resolution to suit the
material being copied, and receive signals from the far end
indicating the level of success of the transmission. Many
Fax machines now offer the ability to transmit an identity
code to the other end. This is useful on an outgoing call
to confirm the identity of the user to whom the document is
about to be sent.

6.5 PSTN CHARACTERISTICS

        The local network is described in some detail in
Chapter 1, and the exchange interfaces are given in Chapter
7. These two sets of considerations give rise to a number
of requirements which must be taken into account when
designing customer terminals for connection to the PSTN.
        In the idle state, current must not be drawn from the
line, since this appears to the line test equipment as
leakage and the line could be classed as faulty. In

practice a maximum of 120 μA per line is permitted, and this can be used for example to power RAM in the terminal. For the same reason earth must not be connected to either leg, except when used as part of a signal to the exchange (such as recall).

In the off-hook state, the exchange feeds 50 volts from either a constant voltage or constant current source. In the former case, the loop resistance of the line, feed bridge and telephone will cause currents of between 19 mA and 90 mA to flow, depending on the length of line. In the latter, current is limited to generally 40 mA or less. When two telephones are off-hook on the same line, the current must divide between them in such a manner as to ensure that both speech circuits are operational, although it is not mandatory to maintain performance to normal limits under these conditions. When a telephone is sending loop disconnect pulses to the exchange, it must not draw more than 500 μA during the break period, otherwise distortion of the pulse will occur at the exchange interface. However, telephones and other terminals must contend with complete breaks in the feed, lasting for up to 200 ms and occurring at any time during the call, including the dialling phase. Terminals in the UK must be insensitive to the line polarity; not only do BT not undertake to supply the negative feed on one particular wire, they do not guarantee that it will remain the same throughout the call. Other administrations are more strict about polarity, and some use a reversal to convey the answer signal to the calling party.

Designers must ensure that terminals do not send unwanted signals to line at power levels which may cause interference with other equipment in the network. Above 50 kHz all signals must be less than -70 dBm. Within the normal audio band (300 - 3.4 kHz) certain frequencies are reserved for signalling systems used in the network and therefore the signal power at these points must be low. However, these power levels are not generated by normal levels of speech. Designers must also ensure that their terminal does not present an imbalance about earth. This helps to reduce potential problems caused by induced longitudinal voltages in the local network. The terminal must also be designed to protect the user, and itself as far as possible, from hazardous voltages that may appear on the network. These take the form of mains voltage, present because of some inadvertent connection, or the much higher voltage of a lightning strike. BT undertake to provide primary protection in the socket by means of a gas arrester discharge tube. This still leaves the terminal needing to protect against 600-1000 volts, and this is normally achieved using a high voltage zener, or a voltage dependent resistor, together with a resistor for limiting the current during the surge.

6.6   TELEX TERMINALS

Standard teleprinters interface with the Telex network

over two wires, one used for transmit data and the other for receive. The signals on each wire consist of pulses of ±80V with an information rate of 50 bit/s. Each character consists of a 5 bit code together with start and stop bits, with the terminal and the network interacting in an asynchronous mode.

National calls are set up using a standard dial at 10 i.p.s. to send pulses into the local exchange, while International exchanges provide registers to respond to keyboard signalling. All teleprinters provide for automatic answering so that they can operate unattended, and when an incoming call is detected the first response is to send an answer back code to confirm the identity of the called terminal. A paper tape punch is provided so that messages can be prepared off-line for subsequent transmission; this facility is useful for long messages, or for international calls where it is important to send the entire message at 50 bit/s without pauses, to minimise the cost of the call. When the teleprinter is being used off-line, an incoming call generates an audible warning so that the operator may choose to suspend tape preparation and accept the incoming call.

Modern forms of Telex terminal are screen based. The operator prepares the message off-line using a VDU and word processing software, and stores it in the internal memory of the terminal. This message can be transmitted at a later time, using unattended operation if required. Features can also be incorporated such as repeat attempt on encountering busy, and broadcast of the same message to multiple destinations.

The 80V interface can be replaced by a signalling system based on 1VF, or Single Channel Voice Frequency techniques. The start/stop data is used to modulate on and off a single frequency tone, one for each direction. This method of signalling puts lower voltages onto the cable network, causes less interference to other services sharing the same cable, and provides better demodulation at the receiving end, leading to fewer wrong characters.

6.7   CURRENT REGULATIONS

As part of the process of 'liberalisation' or 'de-regulation' of telecommunications within the UK, BT no longer retain the monopoly in the supply of customer terminals. However, the networks run by BT are not the same as that run by the CEGB for example; it is perfectly acceptable to connect any electrical appliance to the mains, and the only consideration that has to be given is that of safety to the user. (This is true of the domestic situation, although industrial users must take account of any effect of their loads on power factor). This is because the appliance, in general, is only extracting electricity from the mains. By contrast, a telephone is not only sending signals into the PSTN which it expects to be recognised and acted upon correctly, but also expects to communicate through the PSTN with any other telephone. In order to safeguard the overall performance of networks run by BT, for the benefit of all users, the British Standards Institution have issued a set of standards against which terminals must be tested and approved before authority can be given for their connection to that network. The body currently responsible for carrying out the duties of approving customer terminals is BABT - British Approvals Board for Telecommunications, and is completely independent of BT. Some of the Standards for Telecommunications are given below :-

- BS 6301 - Safety requirements for terminals.
- BS 6305 - Requirements of the PSTN.
- BS 6317 - Requirements for simple telephones to enable correct interworking.
- BS 6320 - Requirements for modems.
- BS 6403 - Requirements for Telex terminals.
- BS 6789 - Requirements for more complicated apparatus connected to the PSTN, including loudspeaking telephones.

6.8   DIGITAL TERMINALS

As described in Chapter 12, all the different telecommunications service networks are coming together into one, known as the ISDN, or Integrated Services Digital Network. Terminals for connection to this network are different from all others, and the interfaces to the ISDN are specified in the CCITT I Series Recommendations. Speech services alone do not economically justify a digital interface at the customer's premises, but it is the ability to provide 64 kbit/s of data in parallel with speech, plus the 16 kbit/s bothway signalling channel availability, that ensures the growth of the ISDN will gain momentum. Nevertheless, it is worth considering the digital telephone.
A digital telephone is defined as having full analogue to digital conversion contained within the basic instrument. The signals conveyed over the pair between the telephone and the exchange are completely digital, superimposed upon the DC power feed. It is not an unreasonable assumption to make that a digital telephone will always be more expensive than an analogue telephone; the digital telephone appears to be the same as an analogue telephone but with additional circuitry. It still requires speech transducers, call alerting device, handset, keypad and hookswitch in addition to the digital CODEC and transmission circuit. There are, however, a number of important differences between a digital telephone and its analogue counterpart :-

- the send and receive levels are totally independent of line length and therefore regulation is not required.

- the telephone is inherently four wire, therefore a hybrid transformer or its electronic equivalent is not required.

- there is no naturally occurring sidetone in the telephone that requires to be eliminated. In order to be in accord with present custom and practice it will be necessary to deliberately introduce some of the transmit analogue audio signal into the receiver. However, this is not to say that sidetone problems have been dispensed with altogether; sidetone may still occur at the first 4/2 wire analogue conversion point in the network.

- the signalling is purely digital and the circuits used to convey keypad information to the exchange are much simpler than for either 10 i.p.s. or MF signalling.

- all signalling is conveyed 'out-of-band' and there is no need for special arrangements for signals such as 'recall'.

- signalling from the exchange to the telephone is as powerful as in the forward direction, and use can be made of this for facilities such as sending the user alpha-numeric messages instead of call progress tones.

- there is no requirement to send ringing current to the telephone. A digital signal is sent from the exchange and the telephone draws an additional DC power which it uses to drive a tone caller.

In addition to the above benefits, the line circuit in the exchange, which by definition forms part of a digital switching local exchange, is simpler and should be cheaper than the equivalent analogue line circuit. There are by way of a contrast, however, a number of problems yet to be satisfactorily overcome :-

- it is not possible to have simple parallel extension telephones if they are digital. Either the analogue portion of each instrument must be extended onto a bus or a digital circuit must be used to multiplex the telephones together.

- there are certain difficulties in signalling digital information in both directions, at an information bit rate of 144 kbit/s, over the local cable network. The technique known as burst mode is the simplest to realise: the exchange and the telephone use the pair alternately in a time-shared mode. The present achievable range is limited to about 3 km. As an alternative, an echo cancelling technique can be used and this in theory gives a greater range of line length, but in doing so uses some complex analogue and digital circuitry.

- binary digits in the local network may cause some interference to other services in the cable.

- some cable pairs may not be suitable for conveying binary signals.

The main reason for the development of the digital telephone is not simply one of satisfying technological progress. When the techniques are perfected the subscriber will be able to make use of a new range of facilities based on the ability of the local network to carry data and a powerful two-way signalling channel, in addition to speech. The standard interface provides in both directions two 64 kbit/s channels plus a 16 kbit/s signalling channel (collectively known as 2B+D). The B channels can be used for speech or data, assigned as the user wishes. It will be common for speech and data to be assigned simultaneously, and the signalling channel will enable them to be delivered to different locations if desired. All those terminals which currently incorporate modems working up to 9.6 kbit/s on the PSTN will have a data capability of 64 kbit/s; this will have a considerable impact on such as Fax machines, both in terms of speed of transmission and quality of the end result. It is to be expected that ISDN connections will lead to a considerable increase in the number and types of terminals providing customers with telecommunication services.

Chapter 7

Switching

Sydney F. Smith

7.1 INTRODUCTION

A large part of the local telecommunication network as it exists today is entirely concerned with the connection of customers' installations to their local telephone exchange. The evolution of that network is therefore necessarily dependent upon the characteristics of the local exchange and upon the evolutionary possibilities of the exchange itself. Of particular concern of course is the nature of those parts of the exchange which provide the interface to the customers' lines which constitute the local network.

In some designs of exchange, the interface may be readily identified with a single functional element or circuit. More commonly, however, the line is switched through to different circuits at different points in the progress of each call.

Those parameters which were important at a particular point in the call will of course have been optimised for the local network at the level which it had reached in its evolution when the system was first designed. Subsequent evolution of the network is then constrained both by this design parameter and others which appear as side-effects which were not significant originally. An example of this can be seen in the bandwidth limitation introduced by the transmission bridge which will be described later in this chapter. The effect of this on the evolution of local network is that an integrated services digital network can only be achieved by replacing or modifying the local exchanges to switch certain types of traffic over paths which do not contain transmission bridges of that sort.

The objective of this chapter is to illustrate this relationship between the evolution of the local telecommunications network and the exchange interfaces.

7.2 EXCHANGE SWITCHING PRINCIPLES

It is clearly uneconomic and unnecessary for every customer or network terminal to have a dedicated connection to all others at all times. Switching is therefore provided to make and break connections as required.

The lines in the local network generally terminate at a Local Exchange (Central Office) which provides the switching function for the local area. The local exchanges are in turn interconnected by junctions directly or via trunk lines through higher order trunk (toll) exchanges (1,2).

To provide the switching function, the exchange requires a switching network, a control system to establish paths through the network and signalling systems to send and receive the data necessary for the control of connections (see Fig.7.1).

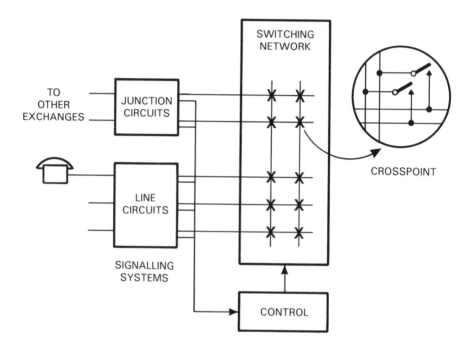

Fig.7.1 Principal elements of a telephone exchange

The switching network comprises an array of single switch elements known as crosspoints. In the older systems these are mechanically operated contacts but a substantial proportion now employ sealed reed relays (1). All these are Analogue systems in which a metallic path is provided through the switching network to convey the speech or other transmission in its original analogue form. The most

modern exchanges employ Digital switching in which
digitally encoded speech (see Chapter 2) is handled by
multiplexed semiconductor crosspoints providing a path
only for the duration of each encoded speech sample (3,4).
The control units in the older exchanges comprise
electromechanical relays in association with mechanical
devices such as rotary selectors and crossbar switches.
The analogue reed relay systems and later digital
electronic systems however employ digital electronic logic
to carry out the path selection process. This generally
takes the form of Stored Program Control (SPC) employing
software controlled processors.
The signalling systems are concerned with the
interchange of call control data between the exchange and
the customers' lines and with other exchanges. The
principal items to be signalled are requests for call,
called line identity typically in the form of dialled
digits, called party answer conditions, charging
information and call termination.
There are two aspects of the signalling systems.
Firstly there are the interfaces to the lines and it is
these which we are mainly concerned with here. There are
also control functions concerned with managing the
encoding and interpretation of signals within the exchange.

7.3 ANALOGUE EXCHANGE INTERFACES

The primary purpose of a telephone exchange is to
provide a speech connection when required between any of
the telephones connected to the exchange. The earliest
manual exchanges did just this and only this. The power
for the microphone was provided by a single cell local
battery at the customer's premises.
To obviate problems associated with battery
maintenance at customer's premises the Central Battery
(C.B.) system was introduced on manual exchanges and is
now universally employed on automatic telephone exchanges.
A relatively large battery at the exchange supplies power
in the form of direct current over the line wires of each
of the telephones connected to the exchange. Because of
the resistance of the wires, the central battery needs a
higher voltage than was provided by the local batteries.
Batteries having nominal voltages from 12V to 60V have
been used. Modern systems generally use 50V controlled
between the limits 46V to 52V. The positive pole of the
exchange battery is connected to earth. The battery is
usually maintained in a fully charged state by floating it
across the output of a mains rectifier unit regulated to
51.5 + 0.5V. Exchange interface circuits are typically
designed to operate on line loop resistances of up to 1500
ohms or exceptionally 2000 ohms (including the telephone
instrument which has a nominal resistance of up to 300
ohms).
The telephone instrument must anyway be designed with
a switch to allow current from the battery to flow only
when the telephone is in use. This switch, known as the

'switch hook' or 'gravity switch', is operated by lifting the handset. The exchange interface for each line incorporates a circuit element in series with the battery connection which detects when the current flows (Fig.7.2).Electromechanical systems generally use a relay for this purpose but electronic systems whether analogue or digital employ the voltage drop across a series resistor to trigger an electronic switch. This 'line' relay is used to light a lamp on manual CB systems or to start the connection process in the case of automatic systems.

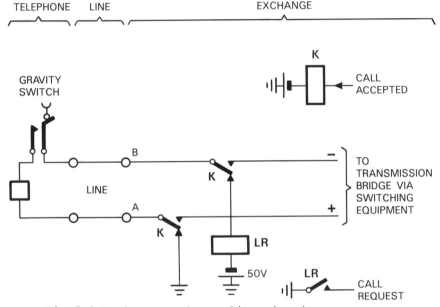

Fig.7.2 Analogue exchange line circuit

The battery and the line relay are permanently connected to the line when the telephone is not in use. In the 'idle' condition however there is no question of crosstalk and no need to maintain a balanced circuit. For economy a simple single element may be used on one wire (− or B wire) with a full earth connected on the other wire (+ or A wire).

When the exchange (operator or automatic equipment) responds to the calling signal (i.e. operation of the line relay),the line relay and earth connection are disconnected and the line is connected to one of a common pool of balanced current feeding circuits. This switching operation may be performed mechanically by the switchboard jack when the plug is inserted or by a 'cut-off' relay (K or CO). The cut-off relay is also operated for calls incoming to the line.

This combination of line and cut-off relay per line constitutes the 'line circuit' and is a common feature of

CB manual, step-by-step (Strowger), crossbar systems and analogue electronic systems.

In some circumstances, earth signalling may be used. An example of this is shared service operation in which each of two customers on the same line have telephones equipped with earthed push buttons for calling. Two line circuits are provided and connected in opposite polarities with the earth return disconnected. Operation of either line relay by the appropriate caller's telephone disconnects the other. The telephone bells are also connected on opposite wires to earth and can be selectively called by the exchange by connecting ringing current to one wire only of the line.

Dial pulses consist of interruptions of the line current. Typical values are 10 pulses per second with a nominal 66ms pulse (break) period. The tolerance on these values depends upon the type of exchange equipment but most will accept pulse speeds from 7 p.p.s to 14 p.p.s. or more with break-to-make ratio of some 50% to 80%. To avoid interference on other lines in the same cable, the relays (A) used to detect dial pulses have two equal coils to maintain a balanced circuit i.e. equal impedance to earth on each wire(Fig.7.3). In some versions of step-by-step system, the line is switched through to a succession of such relays as the call set-up proceeds from stage to stage.

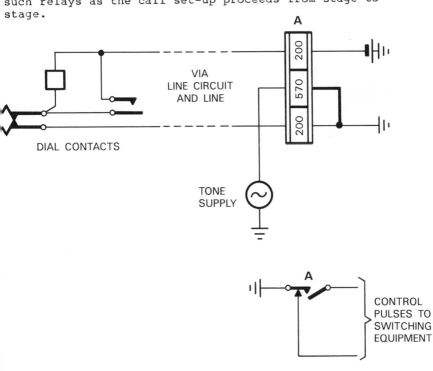

Fig.7.3 Elements of dial pulse and tone circuit

If push button 'dialling' is used at the telephone it may be one of two types. If it is an electronic pulse generating circuit simulating ordinary dial pulses, clearly the exchange interface is the same as for any other telephone. Another type, known as m.f.(multi-frequency) uses oscillators in the instrument to generate a single pulse of a specific two frequencies (see Chapter 6) for each digit to be signalled. The exchange interface then incorporates a tuned amplifier to which the line is connected during the set-up phase of a call following the reception of the initial calling (off-hook) signal.

At certain stages in call set up, tone signals may be returned to the caller (e.g. dial tone). To maintain the circuit balance, these are induced in the line through a transformer which is often provided by adding a third winding to the two coil pulsing relay.

When the connection is completed from the caller to the called line, the exchange connects alternating current to the called line to ring the bell. This 'ringing current' is generated electronically or by a mains or battery driven rotating machine. The voltage is nominally 75V at a frequency of 25 Hz (formerly 16.7 Hz) in the UK or 20 Hz in some other countries. The ringing supply is interrupted in the familiar cadence typically by cam operated springs attached to the ringing machine. The

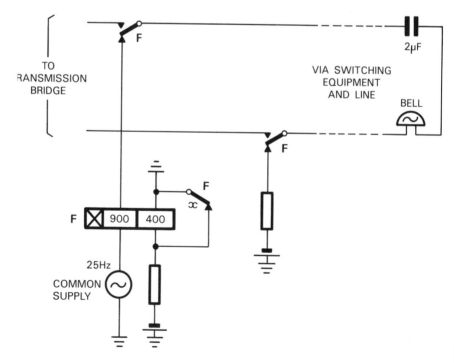

Fig.7.4 Ring trip relay circuit

generator uses earth return and the supply is connected directly to earth during the silent periods of the cadence.

To allow conversation to proceed it is necessary to disconnect the ringing current when the called party lifts the handset. To enable the exchange to distinguish between the two conditions, the telephone has a capacitor in series with the bell to block d.c. The return circuit at the exchange is via the exchange battery and a relay is connected in series with the ringing supply (Fig.7.4). This relay, known as the 'ringing trip' relay (F) is designed to operate to d.c. but not to a.c. alone.

Non-operation is achieved by designing the relay to be too slow in operation to respond to the a.c. between current reversals. The slow operation is produced by means of a short-circuited winding and a 'slug' which consists of a solid copper ring surrounding the core to form a single low resistance short circuit turn. The trip action is provided by an 'early operate' (x) contact of the F relay itself to hold the relay on an independent circuit once it operates. Other contacts of the relay connect the line through to a balanced relay (D) similar to the A relay connected to the caller. This D relay initiates the call charging (metering) process and later detects called party clear (by releasing).

In the conversation phase of the call we now have the calling line connected to the central battery via one balanced relay (A) and the called party connected via another similar relay (D). The inductance of the relays decouples the line from others elsewhere on the exchange deriving their current feed from the same battery. It is then necessary to provide a "bridge" between the A and D relays on the same connection for speech transmission purposes while retaining d.c. isolation between them. The commonest way of achieving this is by means of a capacitor in each wire of the pair (Fig.7.5). The d.c. isolation prevents either line starving the other of microphone feed current and enables separate detection of clearing signals from each party. This combination of two relays plus capacitors is known as a 'transmission bridge'.

For trunk or junction calls (i.e. between customers on different exchanges) it is more usual to use transformer coupling in the transmission bridge. This costs more but has the advantage that low impedance relays can be used which enhance junction signalling performance. It also suppresses longitudinal surges preventing additive effects through a multi-link connection causing signal misoperation.

When the customers replace their handsets to terminate the call, it is the relays in the transmission bridge which respond to the cessation of line current. Their contacts initiate the exchange equipment release process so that the line circuits are once again in control awaiting the next call and the switching equipment is freed for use by other customers wishing to make calls. Each connection must therefore incorporate a transmission bridge at each exchange through which the call has been routed. In the case of a private branch exchange (PBX), the transmission bridge relay is often a single relay in series with the exchange line wires and bridged by capacitors. This means that the relay is controlled by line current from the public exchange independently of the PBX battery so that service is maintained even if there is a local power failure.

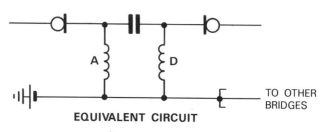

EQUIVALENT CIRCUIT

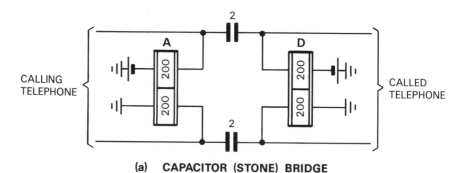

(a) CAPACITOR (STONE) BRIDGE

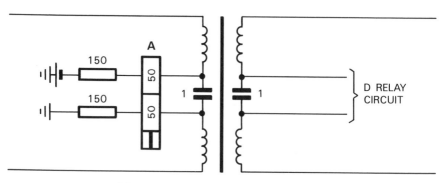

(b) TRANSFORMER (HAYES) BRIDGE

Fig.7.5 Transmission bridges

The transmission bridges are needed only during the conversation phase of the call so their location in the network can be chosen for maximum economy (Fig.7.6). This does however mean that there has to be some switching equipment between the line circuit and the transmission bridge to provide a path compatible with the line itself. In practice this means a two-wire metallic path suitable for d.c. signalling and analogue speech, e.g. Strowger selectors, crossbar switches, or reed relays.

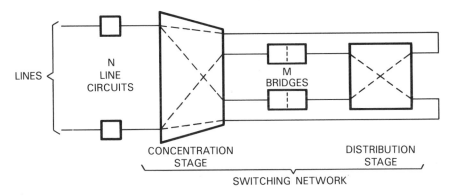

Fig.7.6 Location of transmission bridges in an analogue exchange (N   M)

7.4 INTERFACES FOR DIGITAL EXCHANGES

The availability of reliable high speed electronic switching components has made substantial economies possible in the design of exchange switching networks by employing time division multiplexing. For public exchange systems such as System X the standard method is to convert the analogue speech to digital form by pulse code modulation. The equivalent of the transmission bridge then has to be provided in separate parts at the interface between the digital switch and each line. There is thus a conflict between economy of switch design and the rate of provision of transmission bridges. The effect of reduced switch costs and increased complexity of the interface equipment provided on a per line basis is that the line interface circuits can easily account for more than half the capital cost of a digital telephone exchange. The first digital switching systems solved this by the compromise solution of an analogue concentrator switch with a digital central switch.

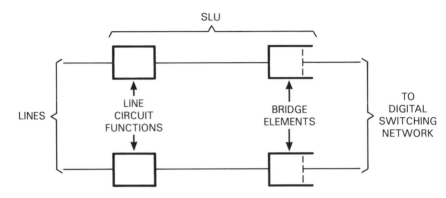

Fig.7.7 Location of transmission bridge elements in a fully digital exchange

Systems now being introduced are fully digital (6) with no analogue concentrator stage and with the transmission bridge function and the line circuit functions provided in the subscriber's line unit (SLU) for each line (Fig.7.7). The SLU also provides the two-to-four wire conversion (hybrid) and analogue-to-digital conversion (Codec) facilities required by the digital switching. The complete range of functions to be provided by the exchange interface to analogue lines at a digital exchange is often described by the acronym BORSCHT as follows (Fig.7.8):-

$\overline{\text{B}}$attery feed
$\overline{\text{O}}$verload protection
$\overline{\text{R}}$inging
$\overline{\text{S}}$upervision (signalling)
$\overline{\text{C}}$odec and filter
$\overline{\text{H}}$ybrid
$\overline{\text{T}}$esting (access for line testing)

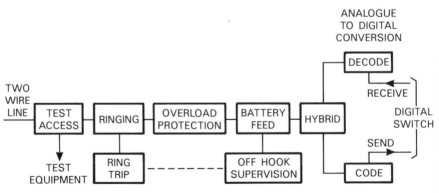

Fig.7.8 Functional elements of a digital exchange line unit

This is currently an area of worldwide development activity (8,9) as semiconductor and telecommunications equipment designers seek to exploit integrated circuit technologies and packaging techniques to make ever smaller, cheaper and more versatile line units.

Single chips incorporating Codec and filter functions for analogue-to-digital and digital-to-analogue conversion have been available commercially for some time. Subscriber line interface (SLIC) chips comprising battery feed, hybrid and supervision functions are now also available from more than one source but they do need external supporting components, such as operational amplifiers and passive balance networks.

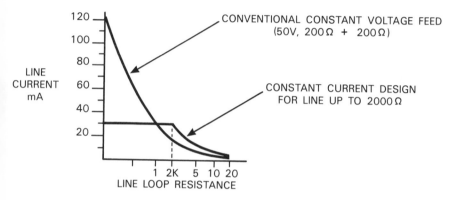

Fig.7.9 Typical battery feed laws

Typically, the battery feed can be made to conform either to the classical constant voltage law, or to a constant current law over a specified range of line lengths (Fig.7.9). One way of achieving this is to use a variable voltage source on the dc-dc convertor principle and controlled by a feedback signal derived by monitoring the line current or voltage. Constant current operation optimises the microphone performance, reduces power dissipation on short lines and reduces variation in impedance between different telephones to assist the design of the two-to-four wire conversion elements. Constant voltage feed has to be used however for telephones employing regulating devices controlled by line current and for interworking with many existing designs of PBX system.

Overload protection in the form of a gas discharge tube is usually provided to some extent at the Main Distribution Frame (MDF) even on electromechanical systems especially if overhead wires are used for any significant part of the line. For the more sensitive electronic components of a digital exchange, faster acting lower voltage secondary protection is usually provided in addition on the SLU itself in the form of protective diodes, fuses and current limiting resistors.

Ringing can either be generated by driving the current feeding circuit at ringing frequency (e.g. 25 Hz) or by switching a common source using relay contacts or high voltage semiconductor switches.

On-hook/off-hook detection for call, answer (ring trip) and clearing (supervision) can be performed by electronic monitoring of changes in line current in the case of constant voltage feed. For constant current feed, the signalling information can be derived from the same source as the control of the variable voltage supply.

The two-to-four wire conversion function can be provided either by a conventional hybrid transformer or by a completely solid state amplifier configuration. The latter has the advantage that it can be more easily adapted to variations in line impedance by varying the value of the balance network components. Currently this involves administrative action to select the appropriate balance value for each line but automatic systems are under development which will adjust the feedback paths by monitoring the signals in the four wire path. Referring to Fig.7.10, the operation is that trans-hybrid signals TH are cancelled by the feedback path FB through the balance network. The use of amplifiers can also be exploited to provide for the different values of 'cross-office' loss in digital local exchanges called for in national transmission plans.

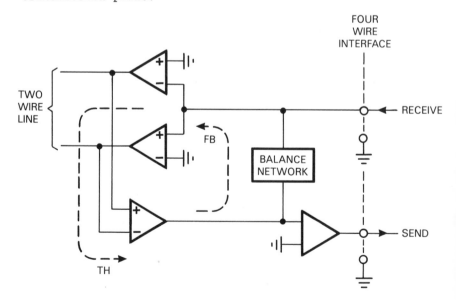

Fig.7.10 Two-to-four wire conversion

The test access function requires a relay if complete
isolation of the line is to be provided during tests
without manual intervention at the MDF. Solid state
switching may be able to provide adequate isolation for
most routine testing but test access is functionally
outside the overload protection so that high voltage
technology is necessary, making it difficult to achieve a
cost effective design at present (10).

7.5 INTERFACES FOR DIGITAL LINES

In parallel with the evolution of telephone exchanges
there have been similar developments in telegraph (Telex)
switching. The power requirements of the customers'
machines however preclude the use of a full central
battery system. A more complex signalling scheme than the
telephony one employing +80V and -80V supplies at both
ends of the line is widely used and a single channel voice
frequency system has now been introduced. The number of
Telex exchanges is substantially less than the number of
telephone exchanges, so it is the latter which are
providing the principal foundation of the evolving
integrated network.
It is this integration of services which brings us to
the latest stage in the evolution of exchange interfaces.
Telephony (speech) services and a variety of text and data
services will share the new common network which will be
completely digital down to the customer's premises. The
telephone then has to be digital as well. In practice this
means providing analogue-to-digital conversion and battery
feed at the customer's premises since the telephone
microphone and receiver are essentially analogue in
function. The line then carries only digital pulses at
typically 80 kbit/s, 144 kbit/s or 2 Mbit/s containing one
or more speech and data channels plus digital control
signalling to replace the present d.c. control signals and
a.c.ringing signals.

7.6 REMOTE SWITCHING UNITS

The simple switching network in Fig.7.1 can establish
a connection between any two of its lines or junctions by
the operation of two crosspoints. All such connections are
made over one of the two links shown by the vertical lines
in the diagram. The traffic from the five external
connections shown is thus concentrated on to the smaller
number of links. In a public telephone exchange we would
expect to have thousands of lines concentrated in groups
on to sets of links which are interconnected by another
network of crosspoints to distribute the calls between the
line groups (Fig.7.6).
In the case of analogue exchanges, the number of
wires between the concentration stage and the distribution
stage is large and it is not usual to provide the stages
in separate locations.

Digital exchanges, however, employ multiplexing so that each concentration group is connected by a small number of multi-channel circuits. This enables subscriber line concentrating units (7) of typically 1024 lines to be located many kilometres away from the main exchange to which they are connected by say, four 30-channel PCM Systems (Fig.7.11).

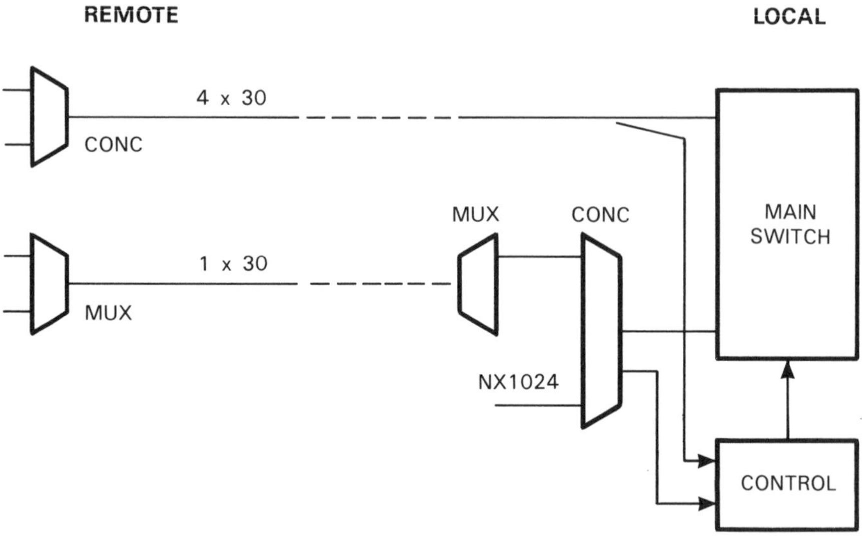

REMOTE   **LOCAL**

Fig.7.11 Remote subscribers' units

For smaller groups of subscribers remote multiplexers can be used without switching concentration.

The multiplexed connections require fewer cable pairs than direct connection by one pair of wires for each subscriber giving substantial savings on line plant costs. The use of remote units also enables much of the switching equipment to be accommodated in buildings outside the expensive city centres. An additional use of concentrators or multiplexers is to provide service to small groups of subscribers needing ISDN facilities in an analogue exchange area.

7.7 CONCLUSION

Exchange interfaces have evolved as an integral part of the local telecommunications network. Most are at present analogue-to-analogue central battery types using largely electromechanical equipment characterised by inductive components, variable impedance and limited bandwidth. A new generation of analogue-to-digital is now going into service characterised by better control of design parameters, controllable current feed and solid state technology. The emergence of digital-to-digital interfaces will accompany the evolution of the network beyond the present speech-only system into an integrated services network.

Further evolution can be expected into the optical transmission broad band service domain.

REFERENCES
1. Smith, S.F., 1978, 'Telephony and Telegraphy', 3rd edition, Oxford University Press, Chapters 3,4,6 and 8.

2. Smith, S.F., 1983, 'Telephone Networks', Chapter 56.2, 'Electronic Engineer's Reference Book', 5th Edition, Butterworth, Edited by F.F.Mazda.

3. Smith, S.F., 1983, 'Automatic Exchange Switching Systems', Chapter 56.3, Ibid.

4. Hughes, C.J., 1986, 'Switching - state-of-the-art', British Telecom Technology Journal, 4, No.1, 5.

5. Haslam, R., 1986, 'New 8/40 subscriber's line circuit for UXD exchanges', Ibid, 20.

6. Ward, R.C., 1985, 'System X: Digital Subscriber Switching System', British Telecommunications Engineering, 3, 241.

7. Rabindrakamur, K., 'The Function and Use of Remote Concentrator Units', Ibid, 245.

8. Apfel, R., et al, 1982 'Signal-processing chips enrich telephone line-card architecture', Electronics, May 5 1982, 113.

9. Potter, A.R., 1981, 'Interfacing to Digital Telephone Exchanges', Microelectronics Journal, 12, No.3.

10. Ames, J.R.W., 1986,'Subscriber Line Interfaces', British Telecommunications Engineering, 4, 203.

Chapter 8

Private switching systems

Mark Trought

8.1. INTRODUCTION

Efficient communication is the cornerstone to any successful business. External communication is necessary to market its product. Internal communication is necessary to ensure the product reaches the market. This is true whatever the product be it services (banks, solicitors etc) or manufactured goods.

Communications come in many forms from voice, through text and data to video. Real time voice is still the most effective and frequently used means of communication. The introduction of the Integrated Services Digital Network (ISDN) has meant that data and text services are becoming more common but, certainly for the forseeable future, voice communication will remain paramount.

Businesses are dependant upon the capabilities of their voice communication systems. There are a number of solutions to the problem of efficient, cost effective voice communications and these solutions have resulted in products that are described as:-

a) Keysystems - covering the market up to 35 lines
b) Small PABXs - from 30 to 100 lines
c) Medium PABXs - 70 to 300 lines
d) Large PABXs - 200 to 4000 lines

Although there are requirements for voice systems larger than this, they are uncommon and are often catered for by networks of PABX's and keysystems of all sizes.

Until the introduction of Stored Program Control techniques, there were user facility differences between keysystems and PABX's. Whilst there were these differences the only question to be asked about the PABX was whether it would be able to connect together the required number of lines.

The introduction of Stored Program Control (SPC) systems meant that facilities previously only associated with keysystems could be realised on a PABX and thus, at the present time, the facility differences between a PABX and a keysystem is negligible. The explosion of features caused by SPC served not only to enhance the users facilities but also the mangement and maintenance facilities of the system.

The   most significant differences between a keysystem and a PABX are now caused by the size which in turn leads to a cost penalty or benefit:-

i)   Keysystems do not (in general) have a dedicated operator position different from the normal keystation.
ii) Most keysystems indicate the status of all other extentions and trunks on all the keystations.
iii) Much of the end user cost of a keysystem is inherant in the cost of the keystation.

PABX's on the other hand:-

i)   Offer a dedicated high function operator's console.
ii) Have sophisticated facilities to avoid having to know the status of extensions and trunks.
iii) Offer a range of terminals from the 'cheap' PSTN type telephone through to sophisticated multi-function electronic sets.

8.2. FEATURES & FACILITIES

8.2.1. Voice User Facilities

The range of features and facilities available to the extension user is both extensive and confusing. These facilities are available to increase the telephone users efficiency by reducing the number of times he makes abortive calls (caused by busy conditions etc.) and increasing the number of successful calls (by using abbreviated dialling). Hopefully, at least for a while, the "feature war" is over, most keysystems and PABXs offer a comprehensive range of user facilities (many of which are rarely used) which cover all normal requirements.
The most common extension facilities are:-

- Enquiry, Transfer, Conference, Intrusion
- Call Back when Free, Call Back when Next Used
- Diversion on Busy, Ring no Reply, Immediate Diversion
- Extension Abbreviated Dialling
- Various Forms of Call Answering (for example Call pick-up groups) etc
- Manager/Secretary Operations
- Privacy,
- Department Call Distribution Systems

Many of the features are able to be allowed or barred to selected users by means of a Class of Service marking.

Together these features increase the efficiency of the bussiness telephone user.

8.2.2 Operator Facilities

Most PABXs have dedicated operator positions. As the operator is the first point of contact that an outside person has with any business, the performance the operator gives in terms of time to answer and efficiency of handling calls is extremely important. The operator will handle a large number of calls per hour and the facilities offered will enable the operator to be more efficient. for this reason, the operator will:-

. Have a display to enable him to know the exact status of the wanted party.
. Be able to queue calls to extension users.
. Have an accurate indication of calls waiting to be answered (in the case where there are a number of operators, calls should be queued in a manner to give the efficient service).
. Be able to intrude into existing calls.
. Be able to release calls from the board to handle subsequent calls, knowing they will return if they are not answered.

Increasingly Direct Dialling Inwards (DDI) is undercutting the importance of the operator. Once the initial contact has been established, subsequent calls can go directly to the person concerned, bypassing the operator.

8.2.3. System Features

Most of the system features are present to reduce the cost running to the owner of the PABX whilst maintaining a high degree of user activity. As such these facilities include:

. Route Optimisation - to ensure maximum use of low cost routes.
. Route Restriction - to prevent use of high cost routes (because route costs change as the time of day, this facility regularly has a time pattern).
. Alternative Routing - to allow high priority users access to more costly routes when low cost routes are fully occupied.

Added to these are:

. Call distribution methods to increase the probability of a call being answered.
. System Abbreviated Dialling Features where commonly used destinations can be called with only four or five digits.
. Emergency operation to handle efficiently times when a number of users simultaneously need to contact the emergency services.
. Direct Dialling Inwards, a method by which individual extensions become part of the national network numbering scheme and can be called directly without the intervention of operators.

8.2.4. Management Features

Along with all the other enhancements, it has become increasingly necessary for the PABX management to be improved. This has happened in the following major areas.

- The ability to move people within the organisation and to control the Features and Restrictions applicable to those users from a relatively easy to use port by means of simple to use commands.
- The ability to allocate the costs of the telephone system to those departments and users making heaviest use of the PABX by accurately recording all calls (internal and external) by time of day, destination, duration etc.
- By measuring the performance of the PABX, most especially in the areaof common resources, to be able to more accurately determine if the PABX is adequately provided with trunks, operators etc to perform its task.

8.2.5. Maintenance Features

The importance of the PABX or keysystem to the business makes it necessary for the system to be operative every working minute of its lifetime. In many cases this is achieved by a high level of redundancy. In the event of any indivdual major component failing, the failure will be detected, the faulty equipment isolated, and its effects limited. Action will then be put in hand to obtain replacement/repair as quickly as possible. To this end most systems continuously check all items of hardware. In the event of failure the system will take the faulty item out of service, switch to using an alternative equivalent piece of equipment if possible, and report the fault. This report may mean alerting the PABX operator so that she can call out the maintainer or reporting the fault directly to the maintainer allowing maintenance to take place, even if the site is unattended.

Another aspect of maintenance is that of the software controlling the exchange; again it is becoming increasingly common for the maintainer to be able to correct faults in the software without ever having to visit the site by means of Remote Access using modem connections.

8.2.6. Data User Features

Computers are becoming more central to the operation of many businesses. Previously, it has been common for terminals to be connected directly to the host computers. But the advent of Data Over Voice (DOVE) techniques and more recently digital transmission has meant that it is possible to use the PABX or keysystem as a means of concentrating terminals onto computers.

Data Over Voice takes advantage of the fact that only a limited bandwidth of the transmission medium between the telephone instrument and the PABX is needed for speech. Higher frequency signals can thus be superimposed and filtered at either end of the connection without affecting the speech.

This has the following advantages:

. Data terminals can be placed wherever there is a telephone without the need for complex additional wiring.
. Concentration from a large number of terminals to a small number of computer ports means that the number of ports on the host computer can be reduced.
. Data terminals can be switched between a number of services allowing users access to many services from a single point.

At the present time, most of the data services are restricted by the capabilities of the terminals connected. Dumb terminals using V.24 connections are usually connected to the computer for long periods of time. This reduces the effectiveness of the concentration aspect of the PABX. As more and more business computers reach the desk more off-line processing will be possible. Thus the holding times will be reduced and the concentration benefits increased. These holding times will also be affected by the data rate from terminal to computer. Currently maximum rates are about 9.6Kb/s. Digital techniques will increase this to 64kb/s.

8.3. System Overview

The major components of any switching system are shown, in outline in fig 8.1. The use of SPC has made the system software equally as important as the system hardware. The major subdivision of the system can be said to be:

1. The Hardware to interface to the users and to switch the calls.
2. The software to control the hardware and to give the users the features and facilities they have come to expect.

8.3.1. System Hardware

8.3.1.1. The Switch Matrix   All modern large PABXs and most keysystems use digital switching. The relative cost of digital transmission is offset by the compact nature of the switch. Most digital transmission and switching is based upon the CCITT pulse Code Modulation (PCM) technique using A-Law (in Europe) or u-Law (in the USA).
PCM encodes the voice bandwidth (300hz to 3.4khz) into 8 bits encoded every 125 micro seconds. This therefore requires a data rate of 64kb/s in each direction for a speech connection.

There are two different approaches used to supply this bandwidth one is a "Port" Switch and the other a "Bus" Switch.
With a Port Switch, every device on the switch is allocated a "Port" and allowed to transmit or receive 64kb/s into the switch. The switch matrix is a centralised device and every interface must be cabled individually to the switch. This means that the switch matrix must be able to offer a bandwidth equal to the number of ports multiplied by 64kb/s.

In other words for a 32 port system it must offer 2.048mb/s,
for a 64 port system 4.096mb/s,
for a 512 port system 32.768mb/s.

Such a system, whilst it can always guarantee a connection from
any port to any other port obviously very quickly requires a very
large switch matrix and many cables.

Using the Bus switch, every interface port is connected to a Time
Division Bus and "Slots" on that bus allocated on demand. The bus
itself allows the interfaces to be connected along its length, thus
distributing the switch matrix and reducing the cabling needs. The
average extension user, uses his telephone for approximately 20% of
the time. Thus about five times as many extensions can be supported
with the same switch matrix as would be possible with a Port switch
running at the same rate.

Each type of switch has its own advantages, Port switches are
common on keysystems and small PABXs where the cost of the switch
matrix is small because there are only a few ports required. Bus
architectures are more common in large PABXs where the switch
bandwidth required would otherwise be extremely large.

Fig 8.1: COMPONENTS OF A SWITCHING SYSTEM

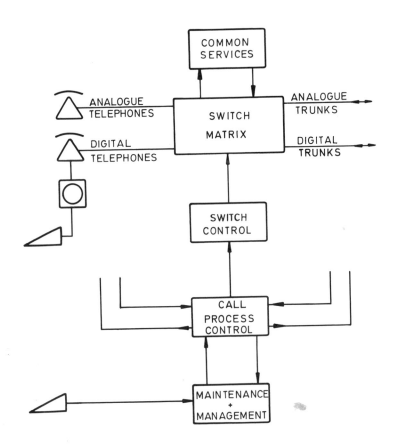

8.3.1. System Hardware – The Control Processors

The ability of the system to perform the user facilities depends entirely upon the successful choice of the processors to control the hardware. Depending upon the maximum system size, the system will be expected to handle between 100 and 100,000 call attempts per hour and the memory requirements vary from 32kilobytes up to many megabytes.

There are a large number of processors available to perform the control from 8 bit micro processors with restricted memory capability through to the newer 32 bit micros capable of addressing gigabytes of memory. Each processor has a different throughput capability.

As the processor is the key to the success of the communications system, and as the processor is not a simple thing to change, it is important to choose a processor with both the necessary speed and addressing capability demanded by the system size, but it is also important to choose one which is able to be supported over many years.

To handle this wide range as efficiently as possible, in many PABXs there are a number of control processors each with their own software dedicated to the different functions. These processors are associated with the interfaces needed to connect the users to the switch thus as the number of hardware interfaces increases, so does the number of processors to handle the load. Thus the critical aspect of the system is more the ability of these processors to effectively communicate with each other than their individual processing power.

The inter-processor communications requirement are affected by the processing architecture chosen. Master/slave relationships create heavy demands on communications as do multi-processor systems. The best solution appears to be multiple processing where processors are largely autonomous with inter-processor communication kept to a minimum.

The memory requirements of the systems have increased because more facilities are expected to be available. Abbreviated Dialling, Route Restriction on time-of-day basic etc., all use up large amounts of data storage.

8.3.1.3. System Hardware – Interfaces

The majority of users are connected through analogue interfaces. In order to minimise the number of these interfaces, the nature of these has changed. Firstly they are now largely software controlled to allow the same hardware to be used for many different purposes and secondly they have been designed to use as little power as possible. The present range of PABXs and keysystems are much more energy efficient than those of the early 1970s although their power consumption is higher than that of Stowger technology.

Digital interfaces are becoming more common. The development of international standards for the ISDN has meant that Digital Trunks and Digital Extensions are now a reality. These have increased the packing density of trunk channels offering 30 transmission channels on a single card. With analogue trunks, an equivalent number of channels would require approximately 12 card positions.

8.3.1.4. System Hardware – Common Services   The use of digital technology and microprocessors has also affected the common services of a PABX. Tone Generation and detection is now through a direct digital representation of the tone which can be quickly re-programmed for different countries. Detection of rotary digits can be performed in software and with digital instruments will not be required. Recorded announcements are also digitally recorded on random access devices giving better overall performance.

Thus again, the requirements for common services needs less physical space in the architecture and so can be more generously provided. This leads to an overall better grade of service to the extension user.

8.3.2. System Software

The software of a communications system can be divided into six major areas:-

1. Interface Scanning:- This is increasingly performed by processors dedicated to the particular interface type. In essence this consists of a means by which any change in the condition of the line (analogue or digital) can be detected, and reporting this change to the Call Processing.

2. Cadencing/Tones:- Again this is largely dedicated to the interface and consists of controlling the ringing cadence and tone cadencing to the user.

3. Call Processing:- This is the heart of the system software. Most systems are designed as "State/Event Machines". In other words every time anything happens, that "Event" is recorded against the call and the call moved to a particular "State". Because the number of "Events" that are valid in any "State" is finite and the effects of any "Event" in any "State" deterministie, this structure leads to a very robust software construction, with effective testing possible.

4. Maintenance/Self Test:- Every part of the system from the memory of the processors to the condition of the Trunk Interfaces has to be tested both on a routine basis and on request by the maintainer. Each test is specific to the type of hardware. For example the memory of the processors may be checked by some form of checksum process that ensures that the memory contents are still valid and the Trunk Interfaces tested by the ability to detect dial tone from the Public Network on seizure.
In the event of any equipment being found faulty, it will be isolated and appropriate action taken. In order to decide what action is required, obviously some form of priority has to be allocated to the fault. As different customers attach different importance to the interfaces on the PABX, such priorites are becoming the subject of negotiation between the customer and his maintainer.

The type of maintenance offered now varies from routine visits, through to an automated call-out by the PABX in the event of a fault being detected.

5.   Management/Man Machine Interface:-   The SPC system offers the system manager the ability to accurately tailor the system so that the best use of the system can be made.   The interface between the manager and the system ranges from a imple MF4 telephone with predetermined facility codes, through a dumb terminal connected to a V.24 port with a set of mnemonic inputs and outputs, to a sophisticated microprocessor with command menus.
The latter is becoming more common as the cost of the micro processor falls and the expertise of the system manager reduces.
This is one area where expert systems will be a future development.

6.   Statistics/Analysis:-   It is important to the system owner that it is possible to accurately assess the cost of running the system.   For this reason most systems are provided with some form of Call Logging, ranging from a hard copy printer through to a sophisticated analysis package capable of producing off-line many different types of report.
It is also important to determine if the system is being adequately used overall.   Thus most systems collect traffic statistics detailing queue length, use of trunks etc. Presented either in tabular or graphical form which allow the system manager to make better decisions over the allocation of trunks, operators etc to make best use of his investment.

8.4. TERMINALS

8.4.1. Analogue Terminals

Analogue telephones have long been based upon those developed for and by British Telecom for use on the Public network.
The faster call set-up required by business users has meant that push button (MF4) signalling has been standard and the use of feature phones   also very common.

8.4.2. Digital Terminals

Proprietory digital terminals offering simultaneous voice and data communications at 64kb/s are now a reality.   In general they are targetted at the top-end of the terminals range because of the inherrent cost of such facilities.   Thus they offer:

a)   Interactive displays to indicate the status of the call (eg Diverted, Number In Do Not Disturb.)
b)   Programmable Feature Buttons.

c)   Single Button access to common features (eg Callback, Conference).

d)   Handsfree dialling with verification.

e)   Loudspeech/Monitor.

A simplified schematic for such a terminal is in fig 8.2.   These are all supported by a sophisticated Common Channel Signalling system and a transmission system designed to work over standard telephone wires   so that they can be put "Wherever there is an existingphone". In private networks, the transmission between the terminal and the exchange will uses a burst-mode technique, only suitable for limited range, because the terminals will normally not be more than 2km from the exchange, the technology is available and currently cheaper than echo cancellation.

Fig 8.2:   COMPONENTS OF A DIGITAL TELEPHONE

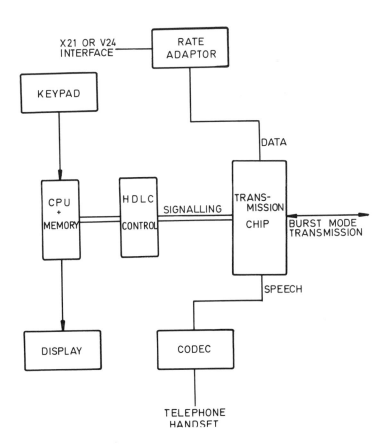

8.5. NETWORKING

Public and Private networks consist of a hierachy of switching systems, Local Exchanges and Group Switching Centres in the public networks, end PABXs and Tandem Switches in Private networks. Pure Tandem exchanges are rare in Private networks as it is normally difficult to justify a switching point with no extensions nearby. Thus the more common use is a sub-tandem where only part of the function of the PABX is to transit trunk calls.

The purpose of any network is to extend the features required by any user into the network. This concept has long been recognised in both the Public and Private areas and has culminated in the current developments of the ISDN. The basic ideas of an ISDN are commonly known and well described in chapter 12. The private switching system plays two roles in an ISDN:-

a)   As an Access System to a Public ISDN.
b)   As an integrated part of a Private ISDN.

In both cases (especially in the latter), the objective of the ISDN is Network Transparency. With a totally transparent network, the network user is not aware of the network topology by any restriction upon the features he may try to execute. This is achieved through the two requirements of:-

1)   Signalling: the user must be able to invoke any feature on any other user.

2)   Transmission: the transmission between any two ports must be the same, regardless of where they are connected.

To understand how ISDN has been achieved, it is necessary to understand the requirements for both Signalling and Transmission, at least in outline.

8.5.1. Signalling

Signalling is required from the user to the switching system and, when PABXs are connected to form a network between the PABXs. Traditionally all signalling has been analogue but these systems are rapidly being replaced by digital systems.

Signalling can be divided into two phases:-

a)   Calling for a register (or recalling the register)
b)   Address signalling to transfer the addressing information into theregister for routing the call.

8.5.1.1. Analogue Signalling – Extensions   The register will orginally be connected by the system detecting the fact that the extension has gone "off hook" and has completed the loop and is now drawing current from the exchange.

When the register has been released, it may be recalled by either:-

a) An Earth on one or other of the signalling paths.
b) A timed break in the signalling paths. The timing of this break may vary from about 50 ms up to 2 seconds.

Most address signalling from the telephone to the PABX uses a tone system known in the UK as MF4. This system uses eight frequencies in two groups of four, and selects one tone from each set, combined to give a single dual tone signal. MF4 is relatively simple to implement and gives a total of 16 distinct tone pairs. MF4 allows about six or seven digits per second to be dialled by the extension user.

This is significantly faster than the more traditional decadic signalling that depended upon accurately timed breaks in the path. This allowed only one to two digits per second to be sent.

8.5.1.2. Analogue Signalling - Networks

Signalling into the public network still uses decadic pulsing on the signalling paths although MF signalling is becoming available.

Whilst all kinds of network signalling exists in private networks, signalling in private networks has largely moved into tone signalling systems because of the additional speed possible. Probably the most sophisticated tone based analogue network signalling system is MF5 and it is worth more than a passing mention.

MF5 is a system based upon two groups of tones (forward tones and backward tones) each group comprising 15 distinct tone pairs. It is a compelled system in that every tone has a corresponding response tone and the next signal cannot be sent until the last signal has been acknowledged.

The forward tones are used to send the address digits and information about the calling party and the backward tones are used to compel the signalling and to give information about the called party (busy, free, etc).

The system is well specified and in different implementations can carry some of the extension user's features out into a network. Its major disadvantage is that, being tone based, it has a maximum information rate of six digits per second and it uses the same transmission path as the speech and so cannot be used at the same time as speech or data.

8.5.1.3. Digital Signalling - Extensions

The most significant developments in PABXs over the past four or five years have probably been in the areas of digital signalling. These have developed from theoretical possibilities into realities in a very short time.

The CCITT ISDN recommendations have a terminal to network structure called "2B+D". This means that there are two 64kb/s transmission channels ('B' Channels) for either voice or data and one 16kb/s signalling channel ('D' Channel). The signalling channel carries all the signalling from the terminal to the network using High Level Data Link Control (HDLC) techniques to ensure validity the of the data. Such a mechanism gives the ability to send and receive data at a very high rate. The signalling between the terminal and the PABX is now limited by the processing capacity of the terminal and the PABX rather than the signalling system.

This has allowed many advantages to the introduced:-

1)   Interactive display phones:-   The information given to the user can now be interactively updated by the PABX, most users will eventually have display phones which will be able to actively track the completion of the call, and to display the identity of the caller etc.

2)   Out of band signalling:-   Much of the interaction between the extension-user and the PABX is able to be done without disconnecting the speech path.

3)   Simultaneous voice and data:-   Because the signalling is out of band, many calls are able to be established simultaneously.

The actual signalling protocols used will vary from manufacturer to manufacturer but all will be based upon the prevailing ISDN access signalling system. The UK is in a somewhat difficult situation at the moment the CCITT Q900 series of recommendations are to be used for full ISDN but the pilot ISDN uses the Digital Access Signalling System (DASS1). Throughout the world such signalling will eventually be according to CCITT recommendations.

8.5.1.4. Digital Signalling – Networks   Digital Network signalling can be further divided into Private Digital Network Signalling and Public Digital Network Access Signalling. In order to simplify and thus accelerate developments, the two systems will be closely related. In the UK this means the Digital Private Network Signalling System (DPNSS) for private networks and the Digital Access Signalling System (DASS2) for 2Mb/s public access. Access to the Public ISDN at lower rates will use the ISDN Q900 series in line with the terminal access.

These systems follow the International Standards Organisation Open Systems Interconnect model at the lowest three layers.

Layer 3	DASS2/DPNSS/Q931	Network Layer
Layer 2	HDLC/Q921	Link Layer
Layer 1	2mb/s Trunks 144k/bs Trunks	Physical Layer

The Physical Layer

The purpose of the physical layer is to transmit the data from one point to another. In Europe all digital networks are interconnected by trunks using CCITT standards at 2.048Mb/s which gives 30 B channels, one D channel and one synchronisation channel. In North America, the standards are different using 1.544Mb/s and 23B+D standard.

The Link Layer

The Link Layer of the signalling system is intended to ensure that messages sent from one end of the link arrive correctly at the other end. To do this they use a combination of error detection and correction. Different mechanisms exist but they all package the data into blocks (known as frames) and perform a checksum on the data sent such that if there is any corruption in the physical layer it can be detected. The frame can then be repeated until it is correctly received or, if it cannot be sent, the signalling equipment can be deemed to be faulty. Chapter 4 gives a fuller description of this technique.

The Network Layer

The Network Layer actually carries the information between the processors of the nodes of the network involved in the call or facility being invoked. There are again many protocols at this level CCITT¢7, CCITT I400 series, DPNSS and DASS to mention but a few. They are all designed to carry information between the exchanges in a network so that the required facilities can be invoked.

DPNSS is probably the most highly featured of these signalling systems allowing many of the facilities commonly expected within a PABX to be offered across a network. The nature of DPNSS with its bias towards a digital physical layer has meant that it has not (as yet) been adopted by keysystems.

The facilities offered (to date) by DPNSS include:-

.Call Back When Free/Next Used
.Diversion (Ring-no-reply, busy, immediate)
.Intrusion
.Three Party Services (enquiry, conference,
 Transfer)
.Centralised Operators and night service

Many PABX manufacturers have implemented DPNSS, with facilities ranging from the very basic, through to full implementations which stretch DPNSS to its limits.

The reason that any digital signalling system is able to extend the complexity of facilities available is twofold.

1. Bandwidth: Common Channel Signalling Systems have access to a complete 64kb/s signalling channel. Because they do not divide it up into a number of fixed bandwidth channels but use messages, which may be on behalf of any channel, the protocols allow signalling rates of over 2.4kb/s per channel (compare this to six digits per second, 24 bits/s for MF5).
2. Out of Band Signalling: Using a non-tranmission channel allows signalling to proceed without affecting the call in rogress.

8.5.2. Transmission

Transmission in analogue networks in the UK is based upon an "incremental loss plan". In that for every switching point, the loss in the transmission path is increased. Digital networks offer loss – free acoustic transmission which means that any two points anywhere on the network can be connected with the same acoustic performance. This introduces other problems in terms of echo and network stability but by careful design, these can be avoided.

Thus we are now at the point where we now have available the necessary building blocks to create an ideal transparent network.

a)   A well defined signalling system capable of invoking the facilities required by a user across many nodes.

b)   A transmission system which means that the user is not aware of the network topology.

8.6. MODERN PRIVATE SWITCHING SYSTEM
8.6.1. PABX

A good example of a modem PABX is the Plessey iSDX (Intregrated Services Digital Exchange). This comes in three versions, the iSDX for the large PABX, iSDX-S for the medium PABX and iSDX-M for the samll PABX.

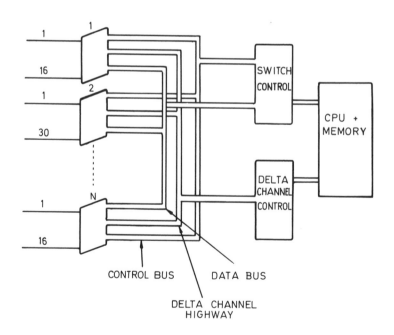

Fig 8.3: STRUCTURE OF THE iSDX

The iSDX (introduced in October 1985) is based upon the iDX which has been in production since January 1983 and has sold nearly 2000 systems.

The switching architecture is a bus architecture capable of supporting approximately 500 simultaneous two way connections and as such can support of the order of 2500 extension ports at 0.2 erlangs.

The processor is a 16 bit minicomputer with an addressing capability of 1 megaword and a processing power of about 1 million instructions per second. The software design is such that the processor is normally about 60-70% utilised allowing 30-40% to be used to handle peaks of traffic. The system is fully redundant with a standby processor and fully duplicated critical equipment.

The system is designed to meet the requirements of customers between 300 and 2000 extensions with over 300 trunks at its maximum size. It is however extremely flexible and is able to be tailored to meet the particular needs of any given customer.

The iSDX-S has a similar architecture using identical software but is designed to meet the requirements of customers from 70 to 300 extensions. As such it has both a reduced switch capacity and a processing power of about 250,000 instructions per second and is not redundant.

The iSDX-M is identical to the iSDX-S in all respects and covers the range from 30 to 80 extensions.

The switching architecture of both is based upon Pulse Code Modulation Time Division Multiplexing A Data Bus carries the data between interfaces and a Control Bus indicates when the Data Bus can be accessed by any given interface.

With the addition of Common Channel Signalling Systems (such as DASS2 and DPNSS), an additonal control bus was added to the iSDX specifically to handle the HDLC signalling on these interfaces. This Delta Channel Highway is controlled and polled using HDLC messages to each interface. The interfaces then terminate the specific level 2 of the external signalling system (DASS2 or DPNSS) but send the level 3 messages through to the main CPU Thus the generalised architecture given in fig 8.1, is in fact realised with two communication highways, one used like a Computer Bus and the other like a Token Polled LAN (fig 8.3.).

The division of the control and switching into these different layers allows new signalling systems to be introduced into the architecture making use of the system features that already exist. This makes it very flexible and easily extended as new signalling systems are introduced.

8.6.2. The Plessey Digital Key System

The Plessey Digital Key System (DKS) was introduced during 1985 and is designed for the keysystem market from 2+6 (2 exchange lines + 6 extensions) up to 10+32. It uses dedicated keystations which give the users many features via single button operation. Like the iSDX it uses PCM A law transmission with the encoding being carried out in the keystation, speech and control signals are digital throughout.

The system uses an 8088processor with a total memory of 36k bytes.

The switching architecture is non-blocking with each port output dedicated to a time-slot on one of the speech highways. Each port can receive from any of the time-slots on any of the speech highways. Thus a bidirectional connection is made by instructing each port to receive from the appropriate dedicated output time-slot for the other party.

A Simplified diagram is given in fig 8.4.

Fig 8.4:   COMPONENTS OF THE DKS

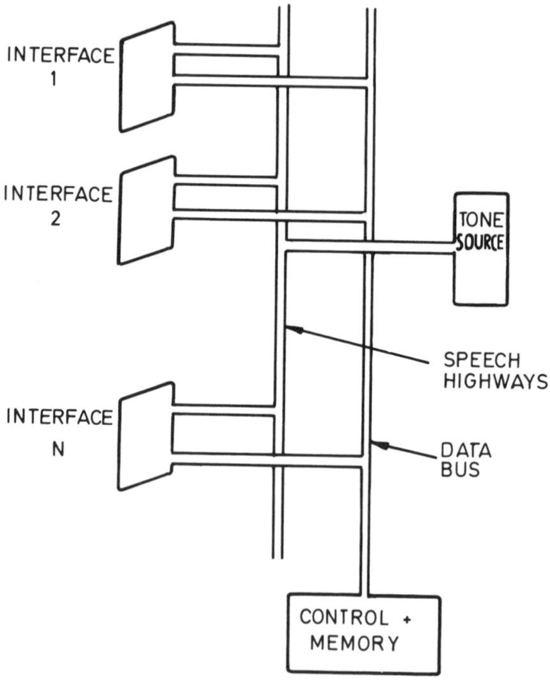

8.7. THE FUTURE

The changes that have taken place over the last few years, direct digital connections between PABX's, Digital transmission to the users desk and Common Channel Signalling have meant that complex networks are now a reality.

The PABX is no longer restricted to voice switching and the availability of simultaneous voice and data transmission from a single terminal removes some of the restrictions previously placed on computer access.   The speeds possible will mean that there will be an explosion in terminals and terminal applications.

Thus, the most significant developments over the next few years will be in the area of data and text applications with Electronic Mail, Packet Switching and Viewdata accessible to all.

Developments on the voice side of the PABX will concentrate in two areas:-

1) Voice Messaging, to allow messages to be left at unattended positons and
2) Voice Recognition/guidance, the ability to make calls without the need for numbers/keypads and help in finding one way through the complex facilities of a PABX.

Developments on the data side of the PABX include:-

1) Improved circuit switch communication for data rates up to 64kb/s from the terminal.
2) Integrated Packet Switching for access to packet switched services.
3) Local Area Network Interfaces.
4) Enhanced Computer Interfaces.
5) Video Switching.

Mixed media developments (Voice annotated text for example) will also develop as a result of the coming together of the two most important means of communication voice and text.

The coming together of all these services will result in a truly Integrated Services Business Communication System designed to speed all the communications of any and every business.

Fig 8.5:   INTEGRATED BUSINESS COMMUNICATION SYSTEMS

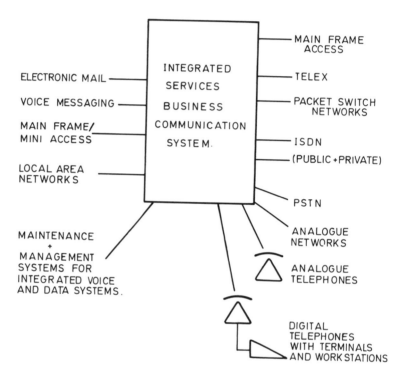

Chapter 9

Commercial aspects

Peter Hughes

Commercial financial aspects are concerned with
return on investment. Any organisation in the private
sector needs a commercial rate of return on the capital
employed in order to pay interest on borrowed capital, to
pay shareholders dividends and to provide capital for the
investment needed to help the business to grow and
improve its operations. This does not imply that consumer
interests take second place. In fact quite the opposite.
Organisations in search of excellence recognise that the
needs of the customer must be satisfied in order that the
business and its offerings are seen to be good value for
money spent.   In this way repeat business is assured.
Where competition exists and customers can obtain the
product or service elsewhere, any failure to satisfy
customers leads to loss of business. The successful
organisation needs to be market driven with a commercial
awareness of the market needs and the nature of the
competition. The commercial enterprise also looks inwards
striving to improve its system and organisation and
reduce cost. The whole process is bound up in a system of
planning starting with corporate plans and the top
management through marketing plans to divisional and
departmental objectives. These are the commercial aspects
that will concern the management of British Telecom   as
the company develops.

9.1   The Transition Period

British Telecom plc is entering a long transition
period as it converts from a large quasi government
organisation in the public sector to a market driven
leader in the private sector. It is possible to sketch a
profile of British Telecom   today and the desired
profile of the future market driven company. The nature
and timescale of the transition period is unclear. So
much depends on the scale and rate of change of
technology as B.T. modernises its system. Changing
attitudes in such a large organisation is a major task in
itself.The external environment is uncertain as Britain

struggles to restore its economy (See 9.1 below).
The following chapter develops the commercial aspects of
marketing, planning and cost reduction within a
regulatory structure and a turbulent environment.

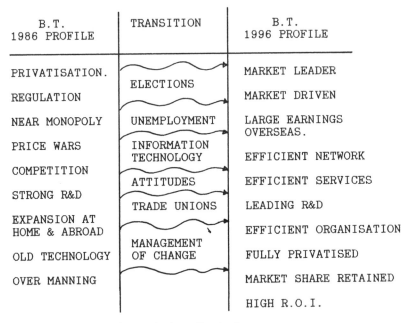

B.T. 1986 PROFILE	TRANSITION	B.T. 1996 PROFILE
PRIVATISATION.	ELECTIONS	MARKET LEADER
REGULATION		MARKET DRIVEN
NEAR MONOPOLY	UNEMPLOYMENT	LARGE EARNINGS OVERSEAS.
PRICE WARS	INFORMATION TECHNOLOGY	EFFICIENT NETWORK
COMPETITION	ATTITUDES	EFFICIENT SERVICES
STRONG R&D	TRADE UNIONS	LEADING R&D
EXPANSION AT HOME & ABROAD	MANAGEMENT OF CHANGE	EFFICIENT ORGANISATION
OLD TECHNOLOGY		FULLY PRIVATISED
OVER MANNING		MARKET SHARE RETAINED
		HIGH R.O.I.

Fig. 9.1 The Transition Period

9.2 LIBERALISATION AND DE-REGULATION OF BRITISH TELECOM

British Telecommunications separated from the Post
Office under 1981 Telecommunications act bringing the 110
year old Postal, Telephone and Telegraph (P.T.T)
combination to an end. The act gave the Government powers
to licence other operators to run telecommunications
systems in competition with British Telecom and British
Telecomm. became the first licensee. The B.T. operating
licence is for a period of 25 years which can be extended
a further ten years; There is to date only three other
licensed national operators. These are: Mercury
Communications, a wholly owned subsidiary of Cable &
Wireless now licensed to run a national telecommunications service in direct competition with British
Telecom The other two licensees operate national
Cellular Radio networks for mobile services. Cellnet is a
consortium of British Telecom and Securicor and
Vodaphone is a consortium of Racal-Millicom.

9.2.1   Privatisation

British Telecommunications was privatised under the 1984 Telecommunications Act. The subsequent flotation of 51% of the shares was successful and five times over subscribed turning B.T. into a public limited company (plc).Prior to the act B.T. was the monopoly provider of telephones and key systems equipment of up to 50 lines. Only larger systems could be bought from independent suppliers. The government set up an independent equipment certification body, The British Approvals Board for Telecommunications (BABT) for the certification of network attachments e.g. telephones, answering machines and private branch exchanges. As   a result there are now 70-100 types of approved telephone instruments available from specialist telephone shops, electrical retailers, department stores and supermarkets. More than 20 rival makes of key telephone systems have been approved by BABT and are being marketed by manufacturers or by specialist dealers and installers. The £2B p.a. investment programme of B.T. makes it a huge market for equipment suppliers. This market , once the sole preserve of U.K. manufacturers is now open to foreign manufacturers.

9.3   OFTEL

Under the 1984 Telecommunications Act B.T. lost its power to grant licences. The government now issues the licences and has set up an Independent Office of Telecommunications (OFTEL) headed by a Director General (DGT) currently Professor Bryan Carsberg. The system of regulation consists of a licensing system operated by the Secretary of State, Department of Trade and Industry. OFTEL will oversee the conduct of licensees and has authority to investigate abuses of monopoly powers and alleged anti-competitive practices. Abusers can be reported to the Monopolies and Mergers Commission (MMC) with its provisions for involving general U.K. competition law.OFTEL has also taken over as a consumer watchdog from the Post Office Users National Council (POUNC). The primary duty of OFTEL is to ensure that telecommunication services are provided throughout the U.K. to meet all reasonable demands.
The subsidiary duties include:-
Promoting the interests of consumers, effective competition, relevant R&D and U.K. interests abroad.
B.T. is required to maintain a range of community services as a condition of its licence. B.T. is allowed to extract a fair contribution of community service costs from operators of interconnected networks. B.T. must show that it is not using its privileged position, as the establised network provider, to gain advantages that are not available to its competitors in terminal apparatus supply. B.T's network telecommunications services may

not cross-subsidise the apparatus business.Any B.T.
manufacturing operation must be a separate subsidiary at
arms length from the rest of the organisation. B.T.must
show that it is not abusing its purchasing power by
requiring suppliers to enter into exclusive dealing
arrangements.

9.3.1   Progress with Regulation

In order for liberalisation and privatisation of B.T.
markets to succeed, it is necessary for OFTEL to exercise
its powers in favour of competition.   B.T. is starting
from an obvious advantage of a massive monopoly of the
telecommunications market.   Mercury   communications , the
only competitor to B.T. for telecommunications services
has installed a trunk telephone network system extending
westwards to Bristol, northwards through Birmingham and
Manchester to Leeds returning through Nottingham,
Coventry and Milton Keynes to the West of London.   There
are microwave links to Cardiff, Swansea, Liverpool,
Edinburgh and Glasgow. The Mercury network is comprised
of optical fibre cables laid alongside British Rail track.
A condition of the licence was that Mercury is allowed to
interconnect with, and use the B.T. network.   B.T. and
Mercury were unable to agree on the terms and conditions
of interconnect. After long deliberation OFTEL have ruled
on the terms and conditions of interconnect.The rulings
establish:-
How much Mercury will pay B.T. for that element of a
telephone call it carries.
Where the two networks will be physically linked and a
timescale for linking the first 36 exchanges.
That Mercury will pay B.T. 50% of the cost of providing
the extra capacity required.
That B.T. will provide full incoming and outgoing
international connections for Mercury.
This ruling by OFTEL is the latest and possibly the most
important to date.   Other rulings include-
(1) OFTEL requested changes to B.T's code of conduct for
consumers in November 1984 and B.T. suitably amended the
code.
(2) OFTEL obtained agreement that B.T. desist from unfair
practices with respect to maintenance, wiring and
additional charges relating to competitors installed PABX
systems.
(3)OFTEL have requested that Public switched telephone
network charges to be on actual rather than average
charge rate.
(4) OFTEL has signalled that B.T. should unbundle
elements of its charging with respect to radio paging.
OFTEL has established itself as the regulatory body in
the telecommunications market and is setting the ground
rules for competition.

9.4 BRITISH TELECOMM.SIZE AND STRUCTURE

British Telecommunications became a public limited company in 1984. The Chairman and Chief Executive is Sir George Jefferson CBE. The company has five divisions :-
  (1) Local Communications Services
  (2) National Networks
  (3) British Telecom International
  (4) British Telecom Enterprises,
  (5) Development and Procurement.
The total group sales turnover in 1985 was £7,653 million with an operating profit of £1,875 million giving a return on net assets of 18.4%. There are a total of 235,178 employees divided amongst the divisions as follows:-

Division	Number of employees at year end
Local Communications Services	197,514
National Networks	10,820
British Telecom International	10,410
British Telecom Enterprises	3,453
Development & Procurement	10,107
Corporate Headquarters	2,874
Total	235,178

B.T. comprises 6000 telephone exchanges including 2100 modern analogue TXE4/4A exchanges and 84 digital electronic exchanges in operation. The balance of exchanges use out-dated Strowger electro-mechanical equipment. There are over 20.5 million exchange lines with copper wire connections to installations, 50,000 kilometres of optical fibre cable has been laid to connect the new digital electronic exchanges. There are 77,000 public payphones and 28.5 million telephones installed.   There are 9000 Cellnet mobile radio phones installed to date.

9.5 THE EXTERNAL ENVIRONMENT

The management of change of B.T. from a huge corporation in the public sector to a market driven organisation in the private sector is very much conditioned by the external environment.The following figure 9.2 shows some of the prevailing influences but in an uncertain future other influences can arise changing the future course of events.

fig 9.2   POLITICS

DECLINE OF MANUFACTURING | OFTEL REGULATIONS

UNEMPLOYMENT ———————(BRITISH TELECOM plc)——— COMPETITION

    SERVICE SECTOR GROWTH | INTERNATIONAL OPPORTUNITY

            INFORMATION TECHNOLOGY

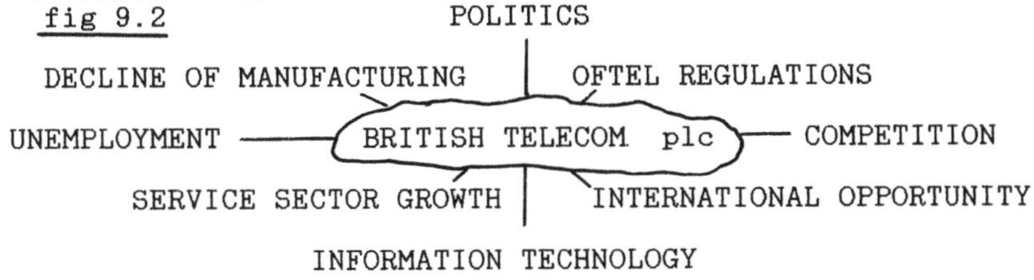

9.5.1 Politics

The long term plans of B.T. are overshadowed by the result of the next election due within 2 years. The policies of the two main political parties and the Alliance must be taken into account in any future plans. If the Conservative party remains in power then privatisation will continue and B.T's transfer into the private sector will be consolidated. If, on the other hand the Labour party is returned as the government then B.T. will almost certainly be re-nationalised along with its competitor Cable & Wireless. Although the leader of the Labour party has indicated that re-nationalisation is not a burning issue, the Labour party view on privatisation is:-
(1) It worsens the lot of the employees.
(2) It is likely to increase unemployment.
(3) It thwarts overall planning of the U.K. economy.
(4) Income goes to institutions and private individuals and not to the State.
The 1983 Labour manifesto referred to the need for firm public control of telecommunications. Trade Unions generally agree with Labour Party policy on privatisation. The Post Office Engineers Union opposed the privatisation of British Telecommunications.
In their 1983 manifesto the Alliance was opposed to taking British Telecom out of the public sector. However, the Alliance is convinced of the need for increased competition and would probably opt to leave British Telecommunications in the private sector.
Politics and economics are tied together in the U.K. economy. Job losses arising from new technology and improved efficiency or from procurement from abroad could become key issues. The increase in charges to ordinary subscribers and any reduction in community services could be criticised by political observers. Discounts to large business users and high profits could draw unfavourable public opinion. The political question is the dominant question over the next ten years for British Telecom plc.

9.5.2 International Opportunity.

B.T. freed to offer it's expertise outside the U.K. is now ready to offer consultancy services abroad and is also prepared to set up and run overseas tele-communications networks. B.T. have entered into a joint venture to develop telecommunications in Malaysia. B.T. has signed a deal with Multitech International that could lead to sales in India. This type of competitive activity is in line with the U.K. governments policy to increase exports of products and services. B.T. recently beat off the challenge from Cable and Wireless for the independent Isle of Man telecommunications network. B.T. has announced that is has received clearance from the Monopolies and

Mergers Commission to acquire a controlling interest in the Canadian Mitel Corporation for $230 millions. Mitel manufacture large and small branch exchanges. B.T. recently purchased U.S. company Dialcom for $13 million with a product compatible with Telecom   Gold micro-computer mailbox service. These and other ventures indicate that British Telecom   is thinking in world terms with respect to business opportunity.

9.5.3   Information Technology

The cost of computer technology has fallen dramatically. At the same time the capacity and speed of computers has increased at an astonishing rate.   In 1965 a silicon chip ready to assemble to a printed circuit board cost about £5.   By increasing the number of circuits on the chip the price fell to about £1.50 by 1975.   Today with say an average of 1500 circuits on a chip the cost has fallen to 0.3p per chip.   In 1975 350,000 silicon chips could store the contents of the complete Encyclopeadia Britannica, today it could be stored on 500 silicon chips. The reliability of computers has improved as fewer connections and less chips are required. Power consumption has been reduced.   Computing speed has increased and access time of computers has decreased with an associated dramatic reduction in data processing cost.   The development of the transputer and beyond that developments in bio-engineering suggest that the computer technology revolution has only just begun.

9.5.3.1   Shift from manufacturing to service industries
There has been a dramatic shift away from manufacturing to service industries in the U.K.   The manpower devoted to handling information has increased to more than 50% of the working population.   Today 40% of the work force is in offices. Office automation will render   using a computer no more terrifying than picking up a telephone. Many desks will have a workstation with a self contained micro-computer.   All work stations and ancillary equipment may be connected on a local area network. Where needed networks and workstations can be connected to a central or national data base by public telecommunications networks. The possibilities extend to home based work stations connected by public telecommunications networks and satellites to networks and data base anywhere in the world.

9.5.3.2   Government initiative in information technology
The Government is taking a positive approach towards information technology.   In the words of the cabinet minister Kenneth Baker; "Information Technology is the fastest developing area of industrial and business activity in the Western world.   Its markets are huge, its application multitudinous   and its potential for increasing efficiency immense.   It will be the engine of

economic growth for at least the rest of the Century".
An awareness of the value of computers and their use by
British industry in becoming more efficient and
competitive lies at the heart of Britain's economic
recovery. The look of the market place is changing.
Business methods and efficiency are being transformed.
New competitors are appearing and a new world of
opportunity is opening. British Telecom and to a
lesser extent Mercury Communications are developing a new
infra-structure of the information society, so that
advanced uses can flourish in telecommunications, in
cable systems and satellites.The Government is backing
the Alvey report to the extent of £350 million to try to
unite the centres of excellence in the United Kingdom -
in industry, universities and research organisations.
The battle for information technology markets is global
with I.B.M. and the Japanese preparing to do battle. The
Japanese Government has allocated £3000 million to
research into fifth generation computers. The huge
investment to convert the public switched telephone
network (PSTN) from a mainly analogue voice medium to a
fully digitalised general data medium for voice, text,
video and numerate services will take place over the next
ten years with British Telecom investing £1-2 billion
per annum. This sort of programme provides a marvellous
opportunity to develop products and services for sale at
home and abroad. In this respect joint ventures with
foreign and/or British concerns could help to secure
entry into International markets. One problem is the
variety of International standards for equipment. The
harmonisation of standards is likely to proceed slowly.
The best chance to gain entrance to a market economy is
probably by inwards investment into that country. The
success of British Telecom plc is crucial to the future
of the British economy. Information technology will play
a large part in shaping the U.K. economy irrespective of
its further development as a service economy, or the
hoped for recovery and return to a manufacturing based
economy. Telecommunications is central to the
exploitation of information technology in the U.K.

9.5.3.3 <u>Technology of the telecomm. network</u> A huge
programme is underway to modernise the U.K. telephone
network. Poor decisions in the past led to a network
that is out of step with present and future requirements.
The out of date analogue network has now been enhanced
with microprocessor controlled TXE4/4A equipment but even
this enhancement leaves the public switching system
slower and far less sophisticated than private
systems.The quality of line is generally suitable for
voice transmission but data transmission can be affected
by the crackles, clicks and fading on the line.
Text and data are transmitted today using modems but the
quality is variable.A typical problem arises in security
systems for banks, shops and offices that suffer false

alarms as a result of poor telecommunications between the premises and the central station monitoring the premises. There is a knock on effect when the under staffed police authority or fire service are obliged to attend false alarms. The problem has got so bad that Police are threatening not to attend a premises following three false alarms. The long term answer is to switch from an analogue to digital system. British Telecom is in the process of installing digital modular System X exchange systems. Co-axial cable is being phased out and replaced with fibre optic cable to connect the exchanges. The plan is to have digital trunk systems in most major cities by the end of 1986. A London overlay network has been laid down by B.T.and can already provide both private circuit switched and leased line digital X-Stream services. A packet switched network is available in most cities linked with European and North American networks via communication satellites. The conversion to an all digital system will be slow despite the huge investment programme by B.T. It is expected that the overall system will be 50% digital by 1990, moving to 74% digital and 26% analogue by 1995 (see figure 9.3 below) Large users e.g. the government, banks, insurance companies, City of London businesses and large companies will have digital trunk facilities far sooner. Once the digital system is in place the stage is set for the information technology and office automation explosion. The telecommunications network will carry improved voice transmission, text, electronic mail, electronic funds transfer, point of sale, credit checking, alarm services, data communications between branch and head offices, graphics and viewphone.The ordinary user is delighted with the new digital phones with their visual display of number, time, and call charge. Once the system is digitalised then connection will be instant and the line quality will be improved. There will be the opportunity for cable T.V. and domestic alarm systems, possibly automatic meter reading and off peak energy control to add to micro-computer inter-connection and viewdata services available now using modems.

9.5.3.4 Martlesham laboratories This huge asset does not appear in the B.T. balance sheet but is of vital importance to the future of British Telecom   The company invests some £200 million annually in Research and Development. If B.T. follows the American AT&T example they will develop their own manufacturing facilities for components and printed circuit boards used in B.T. equipment.Any shortage or failure in supply of critical components could delay the network modernisation.B.T. recently announced a joint venture with Dupont to invest about £100 million over the next four years to mass produce advanced opto-electronic components and lasers for use in telecommunications networks. It is proposed to set up a factory in Ipswich

close to the Martlesham Laboratories. B.T. will continue
to recruit the highest calibre of scientists and
engineers. The new M.Eng. graduates that combine
engineering and entrepreneurship are just the type of
engineers that B.T. need to attract in the future.
Technological innovation is the keystone to future
commercial success.

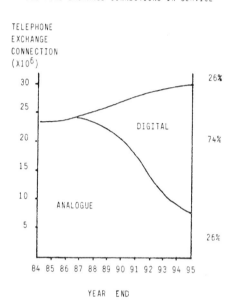

DIGITAL MODERNISATION

TELEPHONE EXCHANGE CONNECTIONS IN SERVICE

fig. 9.3 Digital Modernisation Programme

9.6 BRITISH TELECOM   CORPORATE STRATEGY

Competitors, suppliers and observers have been
surprised at the speed of reaction of B.T. The sleeping
giant woke up with a start and is pursuing an aggresive
marketing policy. There have already been top management
changes within B.T. under the direction of Sir George
Jefferson with a particular emphasis on marketing skills.
Each division is responsible to a main board director
which is the first step in matching responsibility and
accountability.

9.6.1 Core Mission

The policy of B.T. is "to develop British Telecom
into a top class information technology business

operating around the world". The recently liberalised
U.S. company A.T.& T. is to some extent comparable to
British Telecom   and has a similar core mission e.g.
"AT&T's business is the electronic movement and
management of information - in the United States and
around the world". A.T.& T with a turnover of £22,000
million is three times larger than B.T. although B.T. has
the fourth largest network in the world.

9.6.2 Strategic Planning & Marketing

     Part of the transition from a corporation in the
public sector to a public limited company in the private
sector is the need to develop a marketing culture
throughout the company. In a monopoly there can be a
tendency to wait for the business to turn up relying too
much on the latent need for telecommunications to prompt
requests for telephones, telex etc. Lengthy waiting time
for services and delays were a feature of the poor
service before de-nationalisation. Once competition
arrived aimed at the heartland of revenue and profit of
the business then a new urgency became apparent as the
organisation began to question its operations under the
threat of competition.The management were required to ask
questions such as. What segment of the business is under
attack from our new competitors? How attractive is the
alternative service? When will it be available? How
should we react? These and many more questions fall under
the general heading of strategic planning and marketing.
Marketing attempts to obtain accurate and relevant
information about the market segment, the customer needs
and wants, the revenue stream and profitability of the
service in question. Accurate apportionment of costs to
services and products becomes vitally important so that
the contribution to profit and overhead is known for each
product, service or key customer.This information
together with an analysis of the competition is passed to
the strategic planners. It is only with accurate
information that the economists and planners can decide
strategy and price.

Marketing can be defined as:-

"Activities concerned with identifying,   anticipating and
satisfying customer needs profitably".

Or very specifically:-

"Marketing is concerned with selling B.T.telecommunication
network time and selling or renting B.T. supplied equipment
at a profit".

The basic marketing task for B.T. is to regulate the
level, timing and type of demand in a way that will help
the   organisation   achieve   its   growth   and   profit

objectives. Technological forecasting is difficult and prone to error. Although it may be possible to forecast broad trends in information technology and it may be possible to reliably predict when large organisations will adopt new technology. It is much more difficult to predict when the critical mass of small and medium size businesses will start to use available technology.The disappointing start to the B.T. Prestel viewdata service is a case in point. The modernisation of the B.T. telecommunications network may need to be accelerated if information technology takes off at an even faster rate. The British Telecom   strategic and marketing planners need to be alert to the dynamics of the various market segments.

9.6.3 Marketing Planning

B.T. has divided the business into five divisions, one of which is development and procurement. The major market segments are divided up into the following four divisions:-

Local communication services,
National networks,
British Telecom   International,
British Telecom   Enterprises

The top 200 key account customers have been identified within the divisions and product groups. Structuring each market segment and division is the first step in delegating responsibility for action down into the organisation.Product champions can be appointed and any natural entrepreneurs and marketeers arising within the organisation given responsibility and territory. Challenging goals can be set and the organisation becomes goal oriented and market driven.The following figure 9.4 illustrates the type of information and planning process is required for effective planning within a regulatory process.

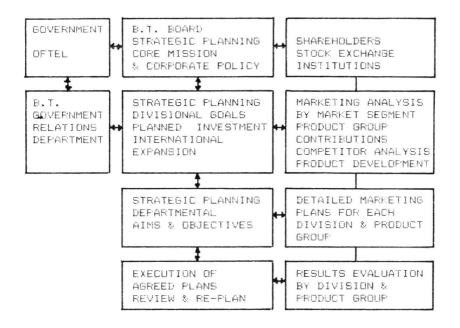

Fig. 9.4   Strategic & Marketing Planning Matrix

9.6.4   Selling v Marketing

The selling and the marketing concept are frequently confused. "Selling focuses on the needs of the seller; marketing on the needs of the buyer"

SELLING CONCEPT

| PRODUCT SERVICES | SELLING & PROMOTING | PROFITS THROUGH SALES VOLUME |

MARKETING CONCEPT

| CUSTOMER NEEDS | INTEGRATED MARKETING | PROFITS THROUGH CUSTOMER SATISFACTION |

Direct selling will be used for key accounts. These are the top 200 purchasers of services or equipment. Top salesmen will attempt to anticipate the needs of these important customers. Marketing started out as an

information service to the sales function. Where key
customers are concerned the salesman is vitally important
in identifying the needs of the customer and closing the
sale.

9.6.4.1 Selection of key markets The first task of top
management is to choose those markets that will be key
markets in terms of size and profitability now and in the
future. The tele-communications market divides into two
main markets:-
(1) Telecommunications services
(2) Telecommunications equipment sales
The two following tables figures 9.5 and 9.6 provide a
forecast of the market segments

B.T. TELECOMMUNICATIONS SERVICES - PRODUCT GROUPS	B.T. REVENUE BY PRODUCT GROUP & PERCENTAGE 1985 - 1990							B.T.MARKET SHARE BY PRODUCT GROUP & PERCENTAGE			MARKET GROWTH PERCENT
	1985 £M	%	CUM %	1990 £M	%	CUM %	DIFF %	1985 %	1990 %	DIFF %	%
BUSINESS CALLS (TRUNK AND LOCAL)	1640	22	22	3100	25	25	-	99	90	-10	+10
INTERNATIONAL SERVICES	1400	18	40	2000	16	41	+2	99	95	-4	+15
RESIDENTIAL CALLS (TRUNK AND LOCAL)	1100	15	55	1800	15	56	-1	99	100	+1	+7
RESIDENTIAL LINE RENTAL	1000	13	68	1500	12	68	-2	100	100	-	+7
APPARATUS RENTAL	950	13	81	650	5	73	-5	90	75	-15	-
BUSINESS LINE RENTAL	390	5	86	600	5	78	-	99	98	-1	+8
PRIVATE CIRCUIT RENTAL	350	5	91	570	5	83	-	100	96	-4	+14
TELEX & OTHER MESSAGING	300	4	95	460	4	87	-	100	98	-2	+8
TELEX RENTAL	100	1	96	170	1	88	-1	85	70	-15	+3
VALUE ADDED SERVICES	140	2	98	390	3	91	+1	70	70	-	+22
CALL OFFICE	100	1	99	180	2	93	+1	100	100	-	+9
ALL MOBILE SERVICES	100	1	100	600	5	98	+3	60	60	-	+40
CABLE TELEVISION REVENUES	-	-		200	2	100	+2	-	25	-	-
TOTAL	7600			12220				97	91	-6	+11

Fig. 9.5 Telecommunications Services Markets

9.6.4.2 Telecom services market It can be seen from
the above table for the telecommunication service market
that the first two segments comprising 40% of the market
are U.K. business calls and International services.
Mercury Communications is the competitor in both these
segments. These segments are expected to grow at between
7% and 10% It is anticipated that B.T. will lose 10% and
5% market share respectively in these two segments by
1990. Residential calls, residential line rental and
apparatus rental are the other major market segments
comprising 80% of the services market by sales turnover.

B.T. is expected to lose 15% of the apparatus rental market share by 1990. The forecast growth rates of some smaller market segments e.g. Mobile services with a growth rate of 40%, value added services with a growth rate of 22% and Private circuit rental with a growth rate of 14% are extremely important sectors which taken together are expected to grow to a combined total of £1.6 billion by 1990.The B.T. share of these markets is forecast to be 60%, 70% and 96% respectively in 1990.

TELECOMMUNICATIONS EQUIPMENT SALES	1985 £M	%	CUM %	1990 £M	%	CUM %	BT MARKET SHARE %		DIFF %	GROWTH RATE %
							1985	1990		
ALL WORKSTATIONS	1000	33	33	2500	47	47	10	25	+15	-20
CENTRAL OFFICE SWITCHING	600	20	53	550	10	57	98	90	-8	-
TRANSMISSION	300	10	63	300	6	63	96	80	-16	-
CUSTOMER PREMISES EQUIPMENT	300	10	73	300	6	69	65	60	-5	-
DATA COMMUNICATIONS	250	8	81	600	11	80	30	35	+5	-10
CABLE	250	8	89	300	6	86	80	80	-	+4
HANDSETS	150	5	94	350	7	93	90	75	-15	-10
MAILBOX	100	3	97	300	6	99	45	35	-10	-25
OTHER	50	2	99	100	1	100	50	48	-2	-15
TOTAL	3000			5300			56	49	-7	-12

Fig. 9.6   Telecommunications Equipment Markets

9.6.4.3 Telecom   Equipment Market The information technology explosion is expected to increase the sales of all workstations in the telecommunications equipment market. B.T. currently holds 10% of this sector although it accounts for 47% of equipment sales for B.T. Vigorous efforts will be made to increase B.T's share to 25% by 1990. Similarly B.T's share of data communications is expected to increase by 5%. The other main equipment market segments show falls in market share for B.T. Share of the Transmission equipment market is expected to drop by 16% whilst Central Office switching and customer premises equipment fall by 8% and 5% respectively. B.T. is expected to reduce share in all other segments by between 2% and 15%. B.T. is more vulnerable in the equipment sales market where many competitors will be attracted into this fast growing market.

9.6.4.4 Competitor analysis The British Telecom competitors are Mercury Communications for telecommunications services and Racal-Millicom for mobile cellular radio communications. The essence of competitor analysis is to know whether to defend market

share by reacting to competitors strategy or whether to
attack the competitors weaknesses. To attack Mercury too
vigorously in the home market could be interpreted as
taking unfair advantage of a monopoly position. To attack
the parent company Cable & Wireless in its overseas
markets may be seen as fair competition. I.B.M. is well
known for its capacity to identify weakness of its
competitors and to expose the weaknesses. Before deciding
on strategy a lot needs to be known about the competitor.
For instance, when will Mercury be able to offer a full
service to potential customers. Will the quality be
satisfactory, or will there be technological and/or
organisational start up problems. It is vital for B.T.
to gauge this before over reacting to Mercury price cuts
on long distance and International call charges.
In the equipment market the problem is more difficult. As
the information technology boom gets underway many more
competitors may be attracted into the market. Large
Japanese and American firms are potential entrants to the
market.The principal competitors   need to be
identified. A thorough and on-going analysis of
competitors and potential competitors is required in the
modern market driven company. Strategy must take into
account both return on investment and competitive
position before decisions can be taken.

9.6.5 Pricing

Pricing is a key issue in the marketing mix. B.T. had
had a policy in the past of subsidising the cost of local
calls from revenues on long distance and international
calls. On the granting of the licence to Mercury
Communications B.T. began to prepare for competition on
long distance and overseas calls by re-balancing its
charges by increasing local call charges and reducing
long distance and international call charges. Further
price increases on local calls are limited to three
percentage points below the Retail Price Index (RPI) -3%.
With low inflation i.e. 3% there is little scope for
price increase on local calls. Mercury Communications
becomes operational in the Summer of 1986 and are
offering price reductions of 17-20% on long distance and
international calls. B.T. will be selective in reducing
prices on long distance calls to match Mercury. An
overall reduction could take £200 million off B.T's
profits. A too large a reduction could bring a reaction
from OFTEL because the aim of the legislation is to
establish a share of the market for Mercury. Indications
are that B.T. will limit price reductions to large
customers with over 20 lines. B.T. may attempt to
increase telephone rental charges to further balance out
charging overall. The following pricing strategy diagram
figure 9.7 indicates the change in B.T. pricing strategy
in response to Mercury's market penetration price
strategy. In the cellular radio market B.T.-Securicor

immediately followed Racal-Millicom in cutting equipment
and installation charges to customers.The key issue in
this market is market share. The price cuts are likely
to drive out the small competitors. B.T. is looking to
retain its 60% of the market and increase its share if
possible. Racal-Millicom are attempting to increase
share. The following pricing strategy diagram figure 9.7
indicates the movement of B.T. Securicor and Racal-
Millicom prices. The volatile price war in the telecom
services segments is likely to be matched in the
equipment market as new technology reduces unit costs.
Once the equilibrium price has been reached firms are
reluctant to reduce price further. Competition takes the
form of advertising and promotion. It took a long time
for IBM to react to price cutting by IBM PC clone
manufacturers. After all, no one was ever fired for
buying I.B.M. equipment.

PRICING STRATEGIES

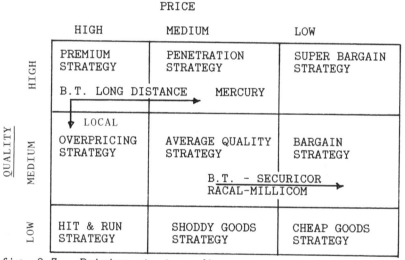

fig. 9.7   Pricing strategy diagram

9.6.6 British Telecom Public Image

    The local communications services have launched a
£160 million programme to modernise the 77,000 pay phones
including vandal resistant equipment and cashless calling
using cards instead of money. There are also a range of
British Telecom shops selling equipment. The telephone
directory has been re-styled and re-launched as the Phone
Book. Accurist are advertising the talking clock service
in a bid to boost calls. There are 430 million telephone
enquiry calls per year. B.T. have introduced a
computerised directory enquiries service to cut enquiry

times from an average of 52 seconds to 30 seconds. B.T.
have taken the needs of the blind, disabled and hard of
hearing members of the public into account in the design
of booths and public telephones. On the downside the
consumer complaint rate has doubled since OFTEL took over
the handling of consumer complaints.Professor Carsberg
the OFTEL DG has criticised B.T. for its teenage
talkabout service that allows up to ten teenagers to have
a shared talkabout call. Irate parents have complained
of bills of £600 in one quarter.
On the business side B.T. are introducing Link Line for
companies who want to pay for their customers calls i.e.
mail order firms, travel firms and hotels. B.T. now offer
simultaneous translation of telephone conversations with
non English speakers or for multi-national conference
calls. B.T. now operate private City of London networks
for fast trading between two points. The line can be set
up in 24 hours. These and many more new services are
aimed at improving and maintaining the company image. A
drop in the consumer complaint rate will signal when
these measures take effect.

9.6.6.1 Advertising The brilliant T.V. advertising
campaign "It's for You-hoo" was a high point, and the
follow up campaign "For creatures large and small make
that call" has not had the same impact. These campaigns
help to boost the image of British Telecom   and are a
vast improvement on the spectacle of giant yellow 'Busby'
birds prancing around London's mainline railway stations
in advertising campaigns of a few years ago. Although it
must be admitted that once seen they are hard to forget.

9.6.7 B.T. as a Marketing Driven Organisation

     In such a large concern it is bound to take years for
the marketing concept to permeate the organisation. The
questions that will decide the marketing effectiveness of
B.T. include:-
(1) Does the management recognise the importance of
designing the company to serve the needs and wants of
chosen markets?
(2) Does management develop different offerings and
marketing plans for different segments of the market?
(3) Does management take a whole marketing system view of
customers, suppliers and competitors, in planning its
business?
(4) Is there a Marketing Director on the main board, and
are the major marketing functions of the company
controlled and integrated at main board level?
(5) Is there good co-operation between Marketing,
Research and Development, Procurement, Operations and
Finance?
(6) Is new product development well organised?
(7) Is marketing information of the required quality in
terms of:- customers, competitors, OFTEL?

(8) Is the sales potential and profitability known for different market segments, customers and product groups?
(9) How cost effective are marketing expenditures e.g. discounts, road shows, T.V. advertising, exhibitions?

9.7 EFFICIENCY & PROCUREMENT CONTRIBUTION TO PROFIT

Profits can come from increased sales, but they can also come from cost reduction with less investment needed. The digitilisation programme means that fewer people will be required to operate the system. B.T. is currently reducing its population by approximately 5000 people a year or 2%. This reduction is by natural wastage. If the need arises to speed up digitilisation or to improve cash flow and profitability there may be a requirement to reduce manning at a faster rate. This could become a key issue in the future for B.T. with its strong trade unions. Procurement offers a good opportunity to reduce costs and improve profitability directly. Liberalisation allowed B.T. to purchase its equipment from anywhere in the world. B.T. policy is to dual source and the company quality assurance survey teams are expert in measuring suppliers capacity to produce products to the required quality. Dual or multiple sourcing provides B.T. with the opportunity to negotiate lower prices. In this period of high investment by B.T. every £1 saved on a contract is a pound on the profits. In 1983/4 only 6% of orders were placed outside the U.K. Liberalisation has attracted foreign manufacturers to the U.K. and this gives a good opportunity to negotiate lower prices with both foreign and U.K. suppliers. The Government is keen to promote inwards investment by Japan and the U.S.A. in order to create jobs. B.T. recently ordered a system "Y" alternative to System "X" from Thorn-Ericsson. The choice of the 5ESS switch for the link programme meant an order for Pye T.M.C. a subsidiary of Philips N.V. These two orders totalled £120M but both Thorn and Pye T.M.C. are long established U.K. companies. Gone are the palmy days of mass manufacture of Strowger 3000 type relays and 746 telephones. British Telecommunications suppliers are now feeling the cold wind of International competition.

REFERENCES

1. Milne, C., 'Operating By Licence' British Telecomm Journal, Spring 1984 25-28

2. Beesley, M.E. 'Progress In U.K. Telecoms Regulations and Competition' L.B.S. Journal Vol.10 Autumn 1985 33-36

3. Nixon, E. Sir 'The Future of Information Technology' L.B.S. Journal Vol.9 Summer 1984 9-16

4.   Pickering, C.   'The Politics of Privatisation of State-Owned Enterprises'   L.B.S. Journal   Vol.9 Summer 1984   27-33

5.   Kotler, P. 'Building A Market Driven Company' L.B.S. Journal   Vol.10   Autumn 1985   19-22

Chapter 10

Local area networks

Steve Wilbur

10.1. INTRODUCTION

By the time significant use was being made of computer systems in the late 1960s and early 1970s, telephone communications were well established both nationally and within organisations. It was therefore not surprising that the early data communications mechanisms relied heavily on modems and the public switched network or leased lines. As computing needs grew so did the need for distribution of access, and with it the need for switching of visual display terminals into computer ports. Typically this was solved by use of switching computers providing a "data exchange" function analagous to the private branch exchange of telephony. However, when the need later arose to share high speed peripherals, such as disks, between a number of dissimilar machines, the switching computer solution was inadequate, unless very carefully engineered. A more general solution was to distribute the switching function along a high speed serial "bus", and with the technology of the late 1970s, data rates of megabits/sec were easily obtained, together with very high switching rates. These Local Area Networks (LANs) developed in a number of ways and some examples will be discussed in detail in this chapter.

The importance of LANs has grown over the last few years as heavier use is made of local computing power on the factory floor and in the office. These machines are rarely isolated, but instead need to coordinate activities, share databases and documents or send electronic mail. In such cases the high data rate is valuable, but only as a means of providing a responsive service when needed. Recently, experiments in using LANs for switching what have hitherto been analogue services, voice and video, have been conducted, and future Integrated Services Local Area Network are a distinct possibility. LANs to support this will need to have considerably higher bandwidth than today's ones. Perhaps, this integration together with the public Integrated Service Data Network will complete a full turn of the wheel from the days when analogue circuits carried voice and data!

Local area networks are generally taken to be digital data networks operating at rates in excess of 1 Mbps over distances from a few metres up to several kilometres. LANs are almost universally serial systems, so that both data and control functions are carried through the same channel/medium, unlike computer busses which generally have dedicated signal paths to resolve access contention. Computer busses have traditionally been parallel systems operating at switching rates of

Megabits/sec over distances of up to about 20 metres. Until recently, wide area network technology was generally limited to about 50 Kbits/sec but the distances covered could range from hundreds of metres to thousands of kilometres. More recently high speed digital data services have become available from PTTs, such as Satstream and Megastream from British Telecom. The latter is potentially capable of digital point-to-point data rates of 2.048 Mbits/sec over the UK mainland. Also emerging, particularly in the US with their cable TV history, are Metropolitan Area Networks, which have many of the characteristics of LANs, but are intended to provide intra-city backbone communications for linking corporate LANs.

Probably the best known LAN work is the development of Ethernet at the Xerox Palo Alto Research Center (PARC). Shoch and Hupp [8] developed the network, and did extensive analyses and measurements. In the late 1970s Xerox re-evaluated it, and re-specified it as a 10Mbps network and a proprietary standard in consort with DEC and Intel. Although this still exists in a revised version [9], the work was also fed into the US IEEE standards body, and has been enhanced in a number of areas as the IEEE 802.3 contention bus LAN [7].

Other work was going on in the mid-1970s, and that of particular UK interest was the Cambridge Ring work at Cambridge University. By the late-1970s several companies were offering Cambridge Ring products. Many of these were being sold to the University sector so the Joint Network Team (SERC/UGC) defined an informal standard called CR82. After having been rewritten in a form suitable for BSI it is now well on its way to becoming a British Standard, and has been submitted to ISO for consideration as an International Standard.

10.2. LAN CHARACTERISTICS

In most LANs, packet switching is used. Thus data is transmitted as a packet containing necessary housekeeping data including synchronisation bits, address bits, user data and check digits. LANs have a fixed address range dependant on implementation ranging from 8 bits (256 nodes) to 48 bits. The latter is intended either to be used in a structured way, ie. broken down into components such as organisation, site, network and host, or to provide a unique address for every machine worldwide. Both destination and source addresses are generally given, but in binary routing networks (q.v.) the destination address is transformed into a route beforehand, and at the destination it may be reversed for reply data. Depending on the network the data size may vary considerably, the range being between about 2 and 2000 bytes typically. Generally, the whole packet is checked using a parity or CRC check. The housekeeping information is very specific to the topology and control technique used.

A characteristic all LANs share is low error rate. Typically this is of the order of 1 in 10^{11} BER, or one bit in error in 3 hours at 10 Mbits/sec. This is poorer than a computer bus but allows considerable simplifications to be made over the protocols used for Wide Area Networks where error rates worse than 1 in 10^5 may sometimes be seen.

There are a variety of topologies which could be used for networks including:

- mesh
- ring
- linear
- star
- tree

A mesh network is built up of links connected to a switching points. The mesh may be regular or just contain those links needed for service plus some redundant links to guard against failures.   With several links at a switching point it is usual to use computers for switching, which generally limit the switching rate to a maximum of several hundred packets per second and tens of kilobits per second. A mesh network has the attractive property of point-to-point links so that the line technology is straighforward.

The limiting case of a mesh is the ring, with each node having one input link and one output link together with the connection to its host machine.   Switching now reduces to deciding whether the passing data stream should be forwarded or passed to the host.   Such a decision can easily be made in one bit time, ie. 100 nsec at 10 Mbits/sec.   The node in such a network performs several functions including switching, timing, and signal regeneration at the line level.   Depending on the actual configuration, this regeneration may be vital, so that power may need to be maintained on part of the node at all times.   Failure of such repeaters or power supplies may seriously reduce the reliability of ring networks, and in critical situations redundancy is provided. Another important point is that LANs are generally designed to have distributed control, ie. there is no single bus arbiter and therefore no single catastrophic point of failure.

The linear network (such as Ethernet) is a simple passive serial bus.   A variety of methods may be used to access the bus but essentially because of its passive nature it is less prone to component failure.   However, because there is a one-to-many transmission pattern the design of the transceivers is more critical than for mesh or ring networks.

Star networks have not been widely used for LANs, probably because of the same switching limitations as mesh networks.   However, there is one exception, being the optical star coupler, which uses a bundle of fibre optic cables fused together as a passive switch.   An input and an output fibre go to each node, so that all transmissions can be heard by all nodes.

Finally, binary routing networks (trees) have been explored for some applications.   Each node in the tree is again required to make only a simple decision on where the data goes.   However, the number of internal connections in a useful sized system is large, making this more appropriate as a backplane technique rather than a distribution technique.

Ring and linear topologies will be further discussed below. For each topology there are a variety of control strategies and these will be described.   Some of the management issues will also be covered, such as how ring reliability might be enhanced for critical applications, and aspects of addressing.

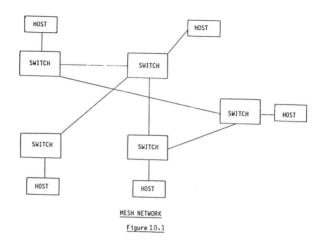

MESH NETWORK

Figure 10.1

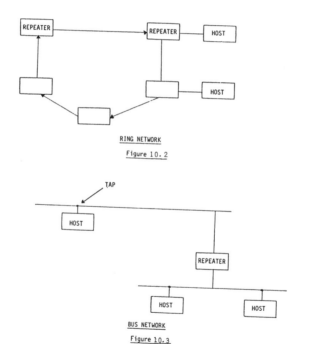

RING NETWORK

Figure 10.2

BUS NETWORK

Figure 10.3

10.3. RING LOCAL AREA NETWORKS

10.3.1 Register Insertion Ring

The register insertion ring was originally designed to switch digitised voice and terminal traffic. Thus, each packet only needs a few bits for data (say 16) and suitable address bits. Each node has a unique address of typically 8 bits, so that the packet size will be 32 bits plus housekeeping bits. In each node is a register of this size which can be loaded in parallel, and then switched in series with the ring. Provided this is done when no data is flowing round the ring or when the end of another packet has passed there will be no confusion. The result is that the node's data is shifted out of the register around the ring and other ring data passes through the register in the meantime. At the receiver a second register is used into which all data passes. At the appropriate moment an address check is made and if the destination in the packet matches the node address, the packet is copied into another buffer (register). When the original packet has done a full circuit the control logic checks the source address against its own and on a match removes the register from the ring.

It will be noted that the packet does a full circuit of the ring. This can be used to pass low level acknowledgement information between recipient and sender, eg. to indicate busy or accepted. Such an acknowledgment facility is extremely useful and worth the apparent 50% loss of bandwidth. (Since most LANs are used at low utilisation levels, eg. 5% peak, this is not usually critical.)

The only way a sender can recognise its transmitted packet is by its source address. If this gets corrupted in transit the node would become deadlocked, and the "zombie" packet would circulate indefinitely. This problem is solved by having a timeout in the node which removes the register from the ring, and adding a "monitor" node in the ring which removes zombies. These can easily be detected by allowing the monitor to write a 1 into a housekeeping bit when the packet first passes it. If the bit is ever seen to be set the packet has to be removed.

In addition to recovering zombies, the monitor can perform initialisation at power-up, ensuring that the ring is empty. Random circulating bits could give rise to spurious reception.

10.3.2 Slotted (Cambridge) Ring

The Cambridge Ring has some similarities with the Register Insertion Ring. However, it is based on the observation that the cable and repeater delays amount to several bits in a typical configuration, at 10 Mbps. Cable delays signals by about 4.5 nsec/metre, ie. 100m delays a signal by 450 nsec or can be thought of as storing 4.5 bits at 10 Mbits/sec. Repeaters with even the simplest line drivers/receivers and switches typically add a further 2.5 bits delay. Thus, a 400m ring with 10 nodes stores about 43 bits and can be thought of as a circulating shift register.

The Cambridge Ring approach divides this shift register into a number of "slots" and a "gap". Each slot contains housekeeping information, addresses and data, amounting to 40 bits. Early in each slot is a full/empty bit. A node wishing to transmit which sees an empty slot passing can mark it full and then fill the remainder of the slot (mini-packet). The recipient seeing a full slot addressed to it can copy the mini-packet into local buffers.

Slots are released by the source node, but rather than detecting the returning slot by address comparisons it can count slots, since there are a constant number of slots circulating in a given ring. The "gap" is therefore important as a datum for this count. It is possible for a transmitter to have been disabled or powered down between transmission and return of a mini-packet, resulting in zombie slots. Once again a monitor is introduced to free up these slots.

The monitor plays several maintenance and management roles in the Cambridge Ring. For very short rings there may be insufficient cable/repeater delay to hold even one slot, so the monitor has a shift register which can be added in series in such cases. Random data is circulated in empty slots and checked on return if the slot is still empty. Error counts are displayed, and if desired reported to a logger on the network. Thus, even if the ring is carrying no user data, latent failures can be detected quickly.

Perhaps one of the important innovations of the Cambridge Ring was the use of the ring itself to diagnose failures. If a ring is broken in two by a cable break, local power dip or certain logic failures, the next node downstream detects this state (actually an illegal signalling state) and generates a "fault mini-packet" giving its own address. This is picked up by the monitor and displayed, so that operators can quickly repair the fault. This mechanism is also used if a slot arrives at a node with an incorrect parity bit. The parity bit is corrected and the slot continues on its way, but a fault mini-packet is later injected into the ring for the monitor and (optional) logging node's attention.

By virtue of its principles, both cable and repeaters must always remain in circuit even when the host computers attached to the nodes are powered down. Thus, power for the repeater section of the node is distributed round the ring itself. In practice, several power supplies are provided to both allow for expansion and to allow for power supply failures.

For data communications purposes a two byte packet size is generally too small. In order to allow simple protocols to be built to transmit kilobyte packets a selection mechanism is included in the receive section of nodes. When the selector is set to N, only mini-packets from source N will be accepted. The values 0 and 255 are reserved for blocking reception from all sources and to allow reception from any node, resp.. An additional response bit is included in the slot to indicate that rejection was due to the select register setting.

The two response bits are coded to indicate one of:

- mini-packet accepted
- receiver busy
- selector set against sender
- destination powered off or non-existent.

The characteristics of both the register insertion ring and the Cambridge Ring are similar. Both offer very low latency transmission, the allocation of bandwidth between nodes being fair. They are ideal for switching synchronous data streams such as voice because of their small packet size and predictable behaviour. Using simple protocols a larger packet can be built up. The small mini-packet size allows low-level flow control, simple hardware interfacing (negligible buffers) but does waste about 50% of the network bandwidth. However, as the data rate is increased larger packets become viable and the penalty is reduced. Thus, future high speed Cambridge Rings may have packet sizes of 40 bytes with only a 10% overhead, ie. user data rates of 60 Mbits/sec on a 70 Mbits/sec ring.

10.3.3   Token Ring

In a quiescent token ring, a single recognisable bit pattern or token continuously circulates. Such a pattern is usually of the order of eight bits in length. When a node is ready to transmit it waits for the token to arrive and changes the last bit to another recognisable pattern, the connector. In this state it is said to hold the token, and can transmit data round the ring. Data up to several kilobytes in length can then be serially transmitted, but only a few bits of this will be on the ring at any time. The recipient recognises the connector and when it sees a packet addressed to it a copy is made into a local buffer. In most token rings the data continues round the ring to the sender which can check that data was not corrupted en route and inspect response bits from the recipient. When the sender has finished transmitting, a token is appended to the end of the transmitted packet.

Several points should be noted. First, because two bit patterns are reserved for the token and connector, these patterns must be bit stuffed when they appear in user data, and are generally avoided in addresses.

In principle, once the token is held very long packets could be transmitted. Normally, a limit is set to how long a token may be held, so that latency is contained within limits and so that a "lost token" failure can be quickly detected. In some token rings a node may only transmit one packet while it holds the token, in others several packets separated by connectors may be sent to any destination provided the time limit is not exceeded.

If the response bit indicating whether a packet is being accepted by the recipient is early in the packet structure, the sender can quickly abort the transmission when refused, so bandwidth is conserved, and the ring latency to other users is decreased.

Broadcast and multicast addressing may be used on token rings, however response bits become meaningless, generally only reflecting the state of the last addressee on the ring. Whilst broadcast addressing is possible on the Cambridge and Register Insertion rings, it is significantly less useful since such small mini-packets are in use, and the flow control for fragment assembly is lost.

The control of token rings is generally completely decentralised, relying on the nodes coming to a consensus in injecting and maintaining just one token in the ring. A typical approach would be for nodes, on detecting loss of token, to insert a new token after a delay proportional to their network address. Detection and removal of duplicate tokens is more difficult since there is little to discriminate between a small ring with one token and a larger ring with two until data is injected. Thus, the startup sequence can be refined as follows. Each node transmits a connector after a delay proportional to its address, immediately followed by a long empty packet. If the packet received is self-addressed, it only remains for that node to insert the token. When an incoming packet arrives, each node stops transmitting, and only if the incoming packet was from a higher addressed node does it attempt to send a connector and long packet again. This process should rapidly converge on a single token ring.

During operation a connector may be changed back to a token by noise with the result that the data returning to a transmitting node is not its own. In such cases the initialisation sequence above must be invoked.

Because larger packets may be used in token rings than in slotted or register insertion rings, the use of the raw bandwidth is potentially higher, eg. 8 bytes of addressing/housekeeping to 20-2000 bytes of data, giving effective data rates of 7-10 Mbits/sec on a 10 Mbits/sec ring. It is generally desirable to have buffers in host memory, but the host memory bus will have its own latency. Thus, a packet buffer in the ring controller or a 100 byte fifo would be used to even out these host latencies. The packet buffer has the disadvantage of reducing the effective data rate by the explicit copying needed, whereas the fifo may occasionally under extreme circumstances be unable to supply/store data at the ring rate. Because the token ring is a packet mode network the access latency is very much higher than the slotted and register insertion rings. Each 2 Kbytes packet might take 2 msec to transmit giving several times this figure as the worst case latency on a large ring with much file transfer traffic.

10.3.4   Packet Protocol

A token ring packet generally consists of the following components:

- destination node address
- source node address
- data bytes
- response bits
- cyclic redundancy check

This is suitable for carrying datagrams between processes, the first few bytes of the data being used to carry sub-addressing information, ie. destination and source port numbers. Datagram protocols may be more complex since the user may want to be able to transmit much larger data chunks, and data streams or fragmentation may need to be provided in these higher level protocols. However, the elements provided by the network are generally those above, except that the response bits are not available for LANs other than rings.

For small-packet rings (slotted, register insertion), a protocol is needed to provide a large packet structure. In the Cambridge Ring two hardware aids to this are provided. This first is the selector register mentioned above. The second are some mini-packet "type bits". Leaving these aside for one moment, a large packet may be provided by a mini-packet stream of the following form and an associated protocol:

- distinctive header
- length of packet
- data bytes
- crc or sumcheck

Initially the receiver sets the selector to 255 (receive all) and when a header arrives the selector is set to the address of the sender. Thus, all other senders are blocked and a dedicated channel is set up between sender and recipient. The next mini-packet would contain the packet length and then the data and check digits would arrive. At the end of the block the selector would again be set to 255.

A few refinements are needed. If the recipient is not ready to receive a packet from the sender, or does not have a buffer of the required length then it can set the select register to zero (receive none) to signal the sender to abort. This will conserve bandwidth. If the recipient gets out of synchronism, it may interpret data as a header. To aid in recovery in such circumstances two type bits have been added to the Cambridge Ring, and one is used to distinguish headers from data. (The other bit is for other protocols, eg. voice.) Thus, a header mini-packet is only recognised if it contains the correct data pattern and the relevant type bit set.

Although rings have a degree of fairness built in, the more persistent senders generally get a larger share of the recipient's services. At the slot level on the Cambridge Ring the hardware limits the retry rate if the destination is busy or unselected. Having just received a packet from a given node, there is a strong chance that a further packet from that node will again be accepted, even though there is congestion at the node. Thus, to be fair, each transmitter should wait a while before transmitting again to the same destination. This is not part of present protocol definitions and occasionally brief lock-outs can be seen in practice. Receivers also need to include a timeout in order to recover from lost mini-packets.

10.4. LINEAR LOCAL AREA NETWORKS

10.4.1 CSMA/CD Networks

The most common control mechanism used on linear LANs is the Carrier Sense Multiple Access with Collision Detection technique (CSMA/CD). It is based on the early work on packet radio done at the University of Hawaii and later refined by Xerox into their early "Ethernet" protocol. The principle is that a node wishing to transmit listens to the cable ("ether") and waits until it is not in use. The node then transmits its data, which propagates along the cable in both directions and is absorbed at the cable terminators. The data transmitted is in the form of packets much like those in the token ring network, so each node inspects the destination address as it passes, and if it is appropriately addressed the data is copied into a local buffer.

However, it is possible for two or more nodes to decide to transmit at around the same time, and for data packets to "collide" on the bus. In the worst case, (ie. two nodes at either end of the network,) it can take twice the cable delay time before a collision is detected at the source. This is termed the "slot time", and for the IEEE 802.3 standard is 512 bits at 10 Mbits/sec (51.2 microsecs). This is the shortest packet which can be transmitted and corresponds to 46 bytes of data plus addressing and other housekeeping bits. When a collision is detected by a transmitter it generates a short burst of "jamming" signal to ensure that the collision is properly detected everywhere and the collision fragment is discarded at all receivers. To prevent repeated collisions after such an event, each participating node backs-off for a random number of slots. The factor is two raised to the power of (a random integer in the range 0 to the retry attempt). Actually, it has been shown to provide good performance in the face of collision if the exponent saturates at 10 for the 11th and subseqent attempts, and a failure is reported if 16 attempts fail.

Although this has been described as a linear network, the standards [1,7,9] use an unrooted tree form with active repeaters between segments. Provided there is only one path between any pair of nodes and the slot time and segment length limits are not exceeded, any configuration can be used. This means that extremely simple systems may have one segment of 500m or less, large building systems may have a "rising main" with spur segments on each floor and large systems may connect buildings by an untapped repeater segment of up to 1500m. The network itself is of co-axial cable with passive taps attached by up to 50m of twisted pair cable to the node and its host.

10.4.2 Token Bus

The token bus attempts to get the best of both worlds, namely a deterministic network access mechanism coupled with a passive medium, ie. a bus. The principle is that a logical ring is formed from a number of the nodes on the bus. Each node determines the address of its predecessor and successor in the IEEE 802.4 draft standard version [6] and the token is generally passed from high addressed nodes to low addressed ones with the exception of the lowest addressed participating node. In the steady state situation a token is passed around and when held a limited amount of data may be transmitted. Some housekeeping is then done to solicit new participants followed by passing the token to the next node in succession. Each stage is, in the absence of noise, deterministic so that minimum system throughput can be calculated in advance.

Establishment and maintenance of the logical ring are important aspects, and are deterministic, although they may involve collisions. The idea of "slot time" as mentioned under CSMA/CD is vital to maintenance functions.

The commonest maintenance function is soliciting new participants, which occurs before passing the token on. A control message is broadcast, called "solicit successor 1" which contains the node's own address and that of its current successor. Any new participant whose address is in this range can then respond in the slot time following the message. If there is silence, the token is passed. If there is a single reply (no collision), the node changes its idea of its predecessor, and the new participant sets up its predecessor/successor values. The token is then passed to the new participant. This address adjustment is possible because each node can hear the control message exchanges. In the case of the lowest addressed node in the ring, the process is slightly different in that a "solicit successor 2" message is sent which waits two slot times for a response. In the first slot time nodes with lower addresses than the present successor are expected to reply, in the second those with higher addresses should reply.

It is, of course, possible for several new participants to reply to such requests, and this can be detected by normal collision detection methods such as signalling errors or check sequence errors. The node then issues a "resolve contention" message and each of the previous respondents choose two bits from their address and wait 0, 1, 2 or 3 slot times before responding again, unless they hear another node responding in the meantime. Thus, a number of the contenders will be eliminated. There may still be contention so the 48 bits of address are used two at a time to resolve it. Because (in error) there may be two nodes with the same address, the final round of contention involves choosing a "random" pair of bits, and bidding in the "resolve contention" slots. If both choose the same value they are both eliminated, otherwise one is added and the other one eliminated. In the worst case 25 resolve contention bids may be needed.

A node may be removed from the ring by either choosing not to respond when a token is passed or, more efficiently, by sending a "set successor" packet to its predecessor while holding the token and then passing it on for the last time.

If a node attempts to pass the token, it also has to check that it has been acted upon. It therefore listens to the network, and if it hears noise or silence (not data) it tries to pass the token a second time. If this fails it assumes the successor has disappeared and attempts to find a successor by sending a "solicit successor 2" packet. If this too fails it tries once more, and after that assumes that it has either gone deaf or the network has broken, and gives up.

At power-up or when the token has got lost, the "inactivity timer" on some nodes will expire. When it does so, the node sends a "claim token" packet with 0, 2, 4 or 6 slot times of padding in the data field based on two bits of its address. After the transmission the node waits one slot time, if there is a response it defers to the other station(s). If there is silence it continues using two bits at a time from the address until it exhausts its address bits or detects another bid after the claim.

An important aspect of the token bus, in common with the token ring, is that it can handle priority data. In the IEEE standard four priority levels are possible. The scheme is based on observing the rate at which the token cycles and only sending low priority traffic when there is adequate bandwidth. Each priority class (except the highest which always gets some bandwidth) is given a target token rotation time. If the actual token rotation time is less than this then the remainder may be used to send this data.

10.5.   OTHER TECHNOLOGIES

10.5.1   Broadband Networks

Rather than using the whole frequency spectrum of the cable, as in baseband communications, broadband networks split it into a number of channels. These channels are not only capable of transmitting data but may also be used for TV relay, security purposes, etc. Such systems are thus often employed where there is a need for a variety of services including data transmission. Although bus-like, the systems differ from those above in that only uni-directional data flow is possible, so two channels are used, one for each direction, with a frequency shifter at the "head-end". Given this, the medium is much like any bus, except that the delay before hearing your own signal, and therefore collisions, is somewhat longer than with baseband systems. Because the medium employs repeaters, long (metropolitan) networks can be built on existing community cable TV systems.

10.5.2   Cheap Networks

There are very many cheap networks operating at data rates up to about 1 Mbit/sec. These are often based on conventional serial I/O chips connected to a common medium. A simple CSMA/CD mechanism can be built in software.

The chips being developed for IEEE 802 and other standards are also capable of driving cheaper co-axial cable with less flexibility of interfacing to provide much cheaper short range local networks.

10.5.3   High Speed Networks

For many years now a 50M bits/sec CSMA/CD network called Hyperchannel has been available for connecting a variety of vendors' mainframes together. Because of slot-time considerations it is limited to a few hundred metres in length, unlike the 2.5km of IEEE 802.3.

There are a number of other developments in progress. In particular there are high speed (70-100 M bits/sec) slotted rings [3], and star-coupler based fibre optic networks under development.

10.6.   MANAGEMENT

In most LAN installations there will be a need to connect LANs together, or to connect them to wide area networks as well as to hosts. It is outside the scope of this chapter to discuss the bridging/gatewaying techniques and also the related table management issues. Instead, this section will concentrate on the low-level management issues.

10.6.1   Ring Reliability

It is often stated that rings are inherently less reliable than bus networks. Whilst this must be true at some level, the MTBF for sizeable rings in practice is of the order of several months, and they often include good diagnostics to provide very low MTTR figures, keeping a very high availability figure. Because of the point-to-point links used, excellent noise immunity, lightning immunity and lack of earthing related problems can be obtained by selectively using fibre optic links.

In cases where exceptionally high availability is required there are a number of enhancements which can be used. The first of these is to provide two rings, all critical resources being attached to both. This is an expensive solution since host interfaces are generally 3-10 times the ring node cost. To gain significant benefits, care must also be taken over "common-mode" faults such as cable for both rings going along the same route, and damage occuring there.

A well-used technique is the "self-healing" ring. Two cables are installed adjacently, and form a pair of contra-rotating rings. The node or repeater logic detects breaks in the ring and attempts to join both together to form one whole ring. With good design, this will cope with both cable breaks and faulty repeaters. The service offered reduces dramatically if two breaks should ever occur.

Another common approach is the "braided ring" where each node is connected to a successor and to that node's successor. The networks can be gracefully degraded within limits as components fail. However, this technique does rely on the braids being sound, and because they are only used under fault conditions, they may have faults themselves which only come to light when they are needed; these are so-called "latent failures".

Another approach is the "switch centre". The ring is formed into a star shape [14], with a reconfiguration panel at the star point. In the case of failures, a given leg can be switched out either manually or automatically. In practice, a single star point is rarely convenient, so a small number of switching points can be introduced as necessary. The links between these points then need to be duplicated, preferably travelling different routes through the building.

10.6.2   Bus Reliability

Bus networks are generally more reliable since they are largely passive. There are however a few failure modes which are important. For good immunity from reflections, cables must be properly terminated. Thus, a cable cut or break, although unlikely, can damage at least one segment of a bus. Active repeaters which fail to transmit data can partition a bus network into several segments. Multiple transmission paths are specifically disallowed in IEEE 802 busses, so standby repeaters have to be switched in and may suffer from latent failures. Good engineering design of the node can limit the extent of most node failures to the node itself. However, it is just possible that a node may attempt to continuously send to the network (jabber). In the IEEE 802.3 specification this must be controlled within 20-150 msec.

10.7.   REFERENCES AND BIBLIOGRAPHY

1.   IT Standards Unit, 1984, "Intercept Recommendations for IT Local Area Networks according to the CSMA/CD Access Method", Dept. of Trade and Industry, UK

2.   Liu MT and Rouse DM, 1984, "A Study of Ring Networks", Ring Technology Local Area Networks, ed. Dallas and Spratt, North Holland.

3.   Temple S, 1984, "The Design of the Cambridge Fast Ring", ibid.

4.   Wilbur SR, 1984, "Aspects of Distributed System Management", ibid.

5.   Falconer RM, 1984, "A Study of Techniques for Enhancing the Reliability of Ring Local Area Networks", ibid.

6.   IEEE, 1982, "Token Passing Bus Access Method and Physical Layer Specifications", Draft Standard IEEE 802.4, Institutute of Electrical and Electronic Engineers Inc, USA.

7.   IEEE, 1984, "CSMA/CD Access Method and Physical Layer Specifications" IEEE 802.3 Standard, Institute of Electrical and Electronic Engineers Inc, USA

8.   Shoch JF and Hupp JA, 1980, "Measured Performance of an Ethernet Local Network", COMM ACM 23, 711-721.

9.   DEC, Intel, Xerox, 1983, "Ethernet LAN: Data Link Layer and Physical Layer Specifications Version 2", DEC, Intel, Xerox Corps, USA

10.   Wilkes MV and Wheeler DJ, 1979, "The Cambridge Digital Communications Ring", Proc. Local Area Networks Symposium, NBS/MITRE Boston, USA

11.   Hopper A and Wheeler DJ, 1979, "Binary Routing Networks," IEEE Trans. on Communications, COM-28, No 10.

2. Bux W, 1981, "Local Area Subnetworks: A Performance Comparison", in Local Networks for Computer Communications, ed. West and Janson, North Holland.

3. Tropper C, 1981, "Local Network Computer Technologies", Academic Press.

4. Saltzer JH, Pogram KT, 1979, "A Star-Shaped Ring Network with High Maintainability", Proc. Local Area Networks Symposium, NBS/MITRE, Boston.

5. Stallings W, 1984, "Local Networks", ACM Computing Surveys, Vol 16, No 1.

Chapter 11

Usable customer interfaces

Ken Eason and Leela Damodaran

11.1 THE EVALUATION OF THE CUSTOMER INTERFACE

The very first telephone users enjoyed the ultimate in
adaptive flexible interfaces. All they had to do was to tap
the microphone and they could use their own spoken, natural
language to ask the system for any service on offer - a 'user
friendly' human switchboard operator did the rest. For the
customer this had many advantages: no constraints on the
dialogue, readily available, explanatory information,
immediate feedback etc.

Since these early days telephony has developed in a rather
unusual way when viewed from the perspective of the consumer.
The range and power of the services it can provide have
become enormous and for most people in the Western World the
telephone has become an indispensable part of daily life.
The technological developments have, however, not only added
to the array of services but have also made it progressively
less necessary to have direct operator intervention. From a
time when the operator managed all calls we have moved
through a period when it was just some calls and services,
e.g. international calls, to a time when there is hardly any
need for operator intervention. Indeed the French by
offering the public direct access to the national telephone
directory are hoping to remove the operator from directory
enquiries. There are many positive features in this
development both for the economics of the business and in
terms of the power that is put directly into the hands of the
consumer. It does mean, however, that successful use of
telecommunications becomes increasingly dependent upon the
ability of the consumer to understand and use the services on
offer. This puts increasing strain on the capability of the
local interface: the means by which the consumer relates to
the network. There is growing evidence that existing
interfaces are becoming bottlenecks which limit the use the
average consumer makes of the services at his or her
disposal. If this is true of the services available today,
it will become a much more serious problem in the future if
plans being laid to add enormously to the telecommunications
services available to the consumer come to fruition. We
could well be heaping technology on technology only to see it
become uneconomic investment because the consumer cannot

cope. We are, in effect, asking the consumer to become capable of operating a very complex array of technical services and, if this is to be successful, the interface by which these services are offered has to be easy to use, must provide good explanations of the services on offer etc. In effect, the technology must perform the duties hitherto performed by the switchboard operator.

In this chapter we shall first review some of the evidence about the use and non-use of present day systems, then we will examine some experimental work on alternatives before identifying the human factor considerations which must be taken into account in systems design if usable access to complex telecommunications services is to be provided.

11.2 THE USABILITY OF FACILITIES IN DIGITAL PABX SYSTEMS

One of the best opportunities we have of studying the likely usability of the wide range of services which may be offered via the telephone to the public in the future, for example by System X, is to study the use of digital Private Automatic Branch Exchanges (PABX's) which already offer to the business community many of these facilities. Although there are variations, there is something of a de facto standard emerging for the provision of the hardware and software of the interface used in these systems. The interface device usually includes a keypad with 12 characters, the numeric characters 0 to 9 and two function keys # and *. The consumer can set up ordinary calls by using the numeric keys but special services or facilities e.g. diverting all calls to another extension, automatically redialling an engaged number etc. are set up using a combination of numeric and function keys. There are a wide array of these special facilities which can mean that the sequence of key depressions necessary to use a specific facility can be quite long, e.g. to store a short code might require *51*33*0123456767 # a total of 17 keys which have to be depressed in the correct sequence. To what extent can the average business user cope with the operation of an interface of this kind? Hannigan and Kerswell (1) report a survey of five advanced PABX's conducted by the HUSAT Research Centre, which provides some evidence to answer this question. A total of 61 users were interviewed on 5 different sites which used different makes and models of PABX. A summary of the pattern of usage of these systems is provided in Figure 11.1.

The cumulative usage record shows that some facilities are very widely used whilst many others are virtually ignored. Five of the facilities account for 48% of usage and 10 for 71%. Only 6 of the 25 facilities on offer were used by more than 50% of the users and for many the frequency of use of

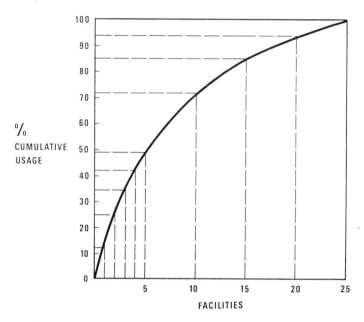

Fig 11.1 Facility Usage in Five Private Computer Based Telephone Exchanges

even this small range of facilities was quite low.   The facilities which were most widely used were 'call pick up', 'call transfer' and 'automatic call diversion'.   These were relatively simple facilities to use and were to some extent obligatory if calls were to be received and routed within an organisation.   Only one facility for dealing with outgoing calls 'system abbreviated dialling' was used by more than 50% of the sample.   This suggests that, where users have discretion as they do when setting up their own calls, they tend to ignore facilities that are complex to use.

It may be argued that most users would not find a lot of the facilities beneficial and indeed we should not expect everything to be relevant to everyone.   However, there are many facilities of which users were not aware or for which they could not state the function, for example, only 8% of the sample knew of a facility called 'executive intrusion'. In other cases it was not the perceived benefit or awareness that was the problem but the complexity of using the facility and the danger of it going wrong.   85% of users knew there was a 'conference' facility but despite its perceived usefulness only 15% had made use of it.

This study shows that there is a low level of usage of facilities even when the facilities are of potential benefit. Other studies, e.g. Craick (2), have obtained similar results for other telecommunication systems.   Eason (3) suggests a framework for explaining user behaviour which proposes that the user makes a series of implicit cost-benefit judgements when using a system.   The benefits are the contribution the facility might make to the tasks the user is currently undertaking and the costs are the time, effort, risk and perhaps financial implications of using it.   One of the problems is that any facility the user has never used or has not used for some time has the additional cost that the user has to find out about it in order to use it.   Over time users tend towards the use of the small set of facilities with which they have become familiar and ignore the rest. If this is the case adding more facilities to be used via an interface which leads to this kind of behaviour may be largely a waste of time and investment.   We have to seek an interface which encourages users to explore all the options which may be of use to them.   In the next section we examine some other options for the interface.

11.3   EXPERIMENTAL STUDIES FOR INTERFACE DESIGN

Many studies have been undertaken to identify optimal forms of man-machine interface but in this section we shall select a sample which is relevant to the design of telecommunication interfaces.   Hannigan and Kerswell (1) report an experimental study which showed some of the ease of use problems of the **#** * protocol.   In experimental use of a system using this form of protocol they found a very high rate of keying errors;   mis-keying, duplication and omissions.   They also found, over an intensive period of trial usage, that subjects did not improve in their performance in the number of errors made or the time they took.   It would seem that what the subjects were required to enter remained meaningless and therefore prone to error. The fact that the **#** and the * changed their function from facility to facility also led to errors and prevented learning since human beings learn by generalising from one circumstance to another.   If there is no consistency this is not possible.   In general this study suggested that it would be very difficult to improve the interface whilst the **#** * was its central feature.

What then are the alternatives?   Man-computer interfaces used in other applications offer many alternatives and it is a question of which of them can be appropriately adapted for telecommunication usage.   Ideally the different commands the user wishes to give the system should have meaningful and consistent names and the user should be given rapid feedback about what the system has understood from the user and what it is doing.   This implies a wider range of functions than the general purpose **#** * and implies a form of feedback to the user.   The series of studies reported by Hannigan and Kerswell (1) included an experimental analysis of an interface based on a series of menus the user could scroll through on a small display and a set of labelled function keys.   The display also presented the numbers dialled by the user and reported the commands it was enacting for the user.

Subjects using this interface did show a learning curve: they improved their understanding of the way the system operated as they tried the various facilities because there was consistency of operation and the activities were meaningful to them.   The subjects were asked to compare this interface with the # * interface and there was a significant preference for the menu-based approach.   There is, however, a cautionary note to be entered about these results: performance overall was better with the #* approach.   The authors attribute this to the unfamiliarity of the menu-based system - the device did not look or behave like a telephone - and there was a degree of learning required before performance started to improve.   The lesson from this study is that there may be better forms of interface but they have to look and behave like the interfaces people know and understand so that they can develop from their existing knowledge base.   The challenge is to create a device which behaves in a way consistent with a telephone but is capable of offering the wide range of facilities in an easy to use, meaningful way.

There are many who believe that the developments in speech processing and in expert systems will provide the best route to an ideal interface for telecommunications.   After all, if telecommunications is primarily for speech communications, what better medium to use for the interface?   Waterworth (4) has provided an excellent review of the potential of voice interactions with machines in the telecommunications context and also identifies many of its problems and limitations. One problem is that sound is a linear medium subject to the constraints of short term memory.   As a result if you hear a menu of options, you have often forgotten the first before you hear the last.

There remain technical problems with voice processing; for example, with the inability of machines to recognise continuous speech from any speaker using an unrestricted vocabulary at normal speaking speed.   Because of these problems we cannot make complete studies of the potential of voice driven forms of interface.   We can however make use of some of the existing, more limited, devices which require people to speak to machines in order to study how people will react to this medium of interaction.   We can, for example, examine how people respond to telephone answering machines when they unexpectedly have to deal not with another person but a machine with which they are expected to interact by voice.   Maskery and Pearce (5) studied how 150 subjects reacted in this situation.   To see what would influence caller behaviour they varied:

1.   the type of voice used to record the announcement
     (female, human, synthetic)

2.   the formality of the announcement (formal, informal)

3.   the amount of instruction to the caller about
     information they should include in their message.

Figure 11.2 summarises the results of this study.

Type of Message   Type of Response

	Good Messages	No Messages	Slam Downs
Informal, instructive	50%	19%	31%
Informal, non-instructive	51%	27%	22%
Formal, instructive	68%	8%	25%
Formal, non-instructive	71%	14%	14%
Formal, instructive, synthetic	64%	6%	31%

Fig. 11.2 Caller Responses to Answerphone Messages

Three types of caller reaction were identified.   These were:

1.   a successful call which resulted in an identifiable
     message being left by the caller (good message)

2.   an unsuccessful call where the caller hung up after the
     announcement message without leaving a message (no
     message)

3.   an unsuccessful call where the caller hung up before the
     end of the announcement message (slam down)

Between 50% and 70% of calls led to good messages which
leaves a very large number of aborted calls.   There were
significantly more successful calls with some kinds of
recorded message.   As figure 11.2 indicates the factors
which seemed to influence user behaviour were:

1.   Formal style announcements result in a higher
     probability of the caller leaving a message.

2.   If the message is directive or instructive there is a
     greater possibility of 'slam downs'.

3.   The synthetic voice led to a higher rate of 'slam downs'
     but not to a lower success rate.

Whilst these results are from an early and unsophisticated
form of voice technology they do suggest that any voice
driven interface has a lot more than technical problems to

solve if it is to be successful; it will have to pay careful
attention to the form and style of its messages if it wishes
people to interact with it. In our everyday dealings with
other human beings we all know that a word out of place or
said in the wrong way can have dramatic effects on the
outcome of the conversation. We will need very sensitive
machines to cope with the sensibilities of the human being.

11.4 CHOOSING A USABLE SYSTEM

There are clearly many research questions to be answered
before we can be sure that the ever widening array of
facilities that can be made available can actually be used by
the consumer. There are indeed some intriguing
philosophical questions which have important consequences,
for example, should we create communication devices which
behave like people (such as voice response systems) or will
this lead consumers to judge systems in human terms and
expect more than the systems can deliver? There is already
a movement that says machines should look and behave like
machines.

Notwithstanding these long term questions there are systems
being developed and purchased today. How can we ensure that
they are effectively used by their consumers? To answer
this question let us imagine that we are the purchaser of a
PABX on behalf of a reasonably sized organisation. The
Managing Director has seen technological innovations go to
waste before because users cannot or will not use them, and
has given us a brief to ensure the system will achieve high
utilisation. What practical advice can be given from the
research that has been done on human responses to these
systems? This advice can be divided into suggestions for
the initial requirements analysis, purchase criteria and
advice on implementation and support procedures.

11.4.1 User Requirements Analysis.

In addition to any technical and cost analysis which may be
undertaken to establish the organisation's requirements of
the new system, there is a need to identify the range of
facilities the users will find of value. Often buyers seek
as many facilities as possible rather than seeking those that
are going to be appropriate. A large array of facilities
makes a complex interface and may make it difficult for the
users to identify and use the facilities they need.
Conducting such a requirements analysis involves consulting
the range of potential users because the nature of people's
responsibilities can change the facilities they may value,
for example, secretaries need and can use different
facilities from managers, the receptionist's needs are not
the same as the Accounts Department, an internal design team
needs different facilities from someone dealing with external
customer enquiries. Identifying the requirements of a
person is not a simple matter of asking them what they want.
The potential user may have no idea what is on offer or what

it could be used for.   It is often necessary to analyse the communication characteristics of the job and to demonstrate available facilities to the users in order to arrive at a considered specification of requirements.   This may seem a lengthy process but is preferable to wasting investment on a system with facilities no one wants and which does not contain the facilities which would really help the business. An analysis of this kind can also help in other ways;   it will show what people have been used to and therefore what kind of learning will be required and it will show what features of the system are going to be important, e.g. what terminology is used within the organisation for communication tasks.

11.4.2   Purchase Requirements.

From a user perspective the important criteria influencing the purchase are the facilities to be provided and the quality of the interface:

1. **Facilities.**   The facilities on offer in a system should be matched with the results of the user requirements study. Several other factors should also be borne in mind. Facilities are unlikely to be in exactly the form required and the provision of a means of tailoring or customising both the facilities and the properties of the interface may be important.   It may also be important to be able to add new facilities unique to the organisation and to be able to adapt and extend the system as the organisation changes over time.

2. **The hardware interface.**   The terminal on the user's desk is the most obvious manifestation of the system and it should meet human factors criteria for ease of use.   Such criteria have been well specified in Cakir et al (6) and Damodaran et al (7).   They cover such characteristics of input devices as the size and shape of keys, the distance between keys, and the force needed to depress them.   They also present the required properties of displays, for example, the size of characters, colours that provide for good legibility, and formatting principles to aid reading.

3. **The software interface.**   Probably the most important aspect of the interface is the messages the system delivers to the user and the messages it can receive from the user which are both under the control of the software.   As a result of the series of studies reported earlier, Hannigan and Kerswell (1) identify the following as the important properties of the software interface:

> *   **The quality of instruction** which has to convey the general principles upon which the system operates, the facilities it contains, their purposes and benefits, and the detailed mode of operation of each part of the system.   The system interface should carry much of this instruction so as to make the interface 'transparent' in

its operation but some of the instructional materials
may be placed in manuals.   An excellent and simple form
of aide memoire is a card or sticky label which can be
attached to the equipment and contains the most vital
codes necessary to get started with important
facilities.

* **Ease of learning** to facilitate the continued growth
of user knowledge, skill and confidence.   Initial
learning is not likely to be comprehensive and the
system will need to encourage the search for and trial
use of other facilities.   In this respect it is
important that the purposes of facilities are obvious to
users and that they can be tried, perhaps 'off line', in
a risk free manner.

* **Meaningful and consistent methods of presenting
facilities.**   In order not to present initial hurdles to
users the system should use terminology and procedures
which closely relate to those with which the user is
already familiar.   If this can be achieved it will
provide external consistency so that the user can
generalise from previous experience.   The system should
also have internal consistency, i.e. all facilities
should utilise the same general principles, so that the
user can generalise from known to unknown facilities.

* **User control.**   The system should allow the users to
set their own pace and find their own way around the
system.   This means for example, that the system should
not pace the user.   In the event of no user response
the system should only default to initial values after
warning the user that this will happen.   Choice should
be available to allow the user to move around the system
but with navigation aids to indicate what is possible at
each stage.

* **Provision of feedback and easy error recovery.**   The
system should never leave the user in any doubt about
the message it has received and the operations it is
performing.   Since it is normal for people to make
errors, the system needs to make these apparent to the
user quickly with guidance on what is required and to
avoid errors leading to calamitous results.

Whilst it may be simpler for the purchaser to assess these
qualities from a demonstration of a system it is easy to
mistake what others will find difficult to use.   It is
therefore desirable that possible systems are tested in user
trials.   It can be a quick and very useful operation to have
a representative group of users examine candidate terminals
and to assess them on the criteria listed above.   It is
important that they actually use the devices and that some
measures of performance with the different facilities are
undertaken.   Without this there is a danger that the user's
preferences will only be influenced by such factors as colour

and style.

11.4.3 Implementation Procedures.

The value of a good system can be lost by a poor
implementation procedure. The following provisions are
necessary if users are to be able to exploit the new
capability with which they have been provided.

1. **Configuring the System.** If good use is to be made of
the system there are a variety of local adjustments to be
made. Decisions have to be taken about the availability of
facilities to members of staff. In some organisations this
has been arranged according to status with the result, for
example, that senior managers get all the facilities and
their secretaries a more limited subset. This is an
excellent way of ensuring many facilities do not get used
because senior managers are often too busy to master unusual
facilities whereas secretaries will treat the mastery of
communications as part of their job. It may also be
appropriate to customise many of the facilities for local use
and where a user would need to load the system with personal
data before it could be used, for example, listing telephone
numbers to be given abbreviated codes, it will be valuable to
offer the user help to do this. One of the barriers to
successful use is the front end effort needed before there is
any benefit and helping users in this way can eliminate this
problem.

2. **Training and Support.** A complex set of facilities
requires knowledge to use well and ideally users will receive
full training with practise in the use of each part of the
system. It is unlikely that this will be possible for more
than a small percentage of the user population. For the
majority of users a short appreciation course may be all that
is possible. If this is the case it is best if it is used
to indicate the general principles of operation, to list the
facilities and their benefits and to introduce supporting
documentation. An alternative is to teach the limited
number of facilities which are likely to be most useful.
Unfortunately this reinforces the tendency to use only a
small subset of facilities. If full, comprehensive training
is not possible it is important that continuing support is
available to users to encourage them to keep exploring the
capability of the system. This can be done by establishing
central liaison staff but another ploy is to fully train one
user in each location and appoint that person as the local
source of support. It has been shown repeatedly that people
seek help from those close at hand and this kind of user
support network can be a very powerful structure in extending
usage of the system.

3. **Organisational Changes.** A common barrier to successful
use of a system is that the organisational changes necessary
to support usage have not been formally agreed. For
example, 'call diversion' and 'call pickup' depend upon a

degree of cooperation amongst members of the organisation.
Establishing who has responsibility for these actions is a
necessary corrollary to use of the system and this is a
management task.   However, management may not immediately
see the need for it since it is a technical system that is
being introduced and insufficient attention is often paid to
these aspects of system implementation.

11.5   CONCLUSIONS

The dangers of introducing sophisticated communication
systems which users cannot use effectively are reduced if the
issues listed above are attended to carefully.   We have
couched the advice in a form suitable for the implementor but
there are obvious implications for the suppliers of these
systems both in terms of the facilities and interfaces
provided and in terms of the support and help given to
customers if they are not to waste the potential of the
product by inadequate implementation procedures.

ACKNOWLEDGEMENTS

We wish to acknowledge the support provided by British
Telecom for the work reported in this chapter and to
recognise the contribution made to the research programme by
the staff of British Telecom and by the following who are our
colleagues in the HUSAT Research Centre: Steve Hannigan,
Anne Clarke, Dave Poulson, Dale Hewitt, Gordon Allison and
Wendy Olphert.

REFERENCES

1.   Hannigan S and Kerswell B, 1986, 'Towards User Friendly
     Terminals' Proceedings of the ISSLS Conference, Japan,
     September.

2.   Craick J, 1985, 'Studies for a Multi-Service Future:
     What do Service Providers Need to Know?' Proceedings of
     the XIth International Symposium on Human Factors in
     Telecommunications, Cesson Sevigne, France, September.

3.   Eason KD, 1985, 'The Man-Machine Interface for the
     Intermittent User', Pergamon Infotech State of the Art
     Review 'The Information Initiative' Pergamon Infotech,
     Maidenhead.

4.   Waterworth J, 1984, 'Interaction with Machines by Voice:
     A Telecommunications Perspective', Behaviour and
     Information Technology, 3, 2 163-178.

5.   Maskery HS and Pearce BG, 1982, 'Telephone Answering
     Machines - An Investigation into User Behaviour'
     Proceedings of the IEE Conference 'Man-Machine Systems',
     Manchester.

6.   Cakir A, Hart D and Stewart TFM, 1980, 'Visual Display
     Terminals' Wiley, Chichester.

7.   Damodaran L, Simpson A and Wilson P, 1980, 'Systems
     Design for People' NCC, Manchester.

Chapter 12

Integrated services digital network

Richard Boulter

12.1 INTRODUCTION

The main features of any network can be categorised into the transmission, switching, signalling and control capabilities of the network. All three of these areas in the Public Switched Telephone Network have evolved in different stages and at different rates. However they have evolved in such a manner as to enable them to converge towards an Integrated Services Digital Network at minimal cost. This chapter briefly reviews this evolution before describing the additional developments carried out to establish British Telecom's (BT) pilot ISDN in the UK. The evolution of this ISDN, based upon the recommendations of the CCITT, are then outlined before finally describing some possible further stages of evolution.

12.2 THE ANALOGUE NETWORK

Until fairly recently the whole of the Public Switched Telephone Network (PSTN) was based upon analogue transmission and switching techniques. The telephone converted the acoustic waves of the speaker into an electrical signal occupying a bandwidth of the order of 4kHz (300-3400Hz). This was transmitted over the telephone network in this form to a remote telephone which then reconverts the electrical signals back to acoustic waves. In what we know as the local network, the go and return paths were provided over the same copper pair since the signal remained in the baseband, whilst in the main network 4-wire techniques were often used.

Switching at the local exchange used space division techniques, originally strowger and later crossbar and reed relay, on the baseband signal in order to concentrate traffic onto junction routes to main network switching centres which were again of the space switching type. Trunk routes then interconnected these main network switching centres and it is on these high capacity routes that we first saw the introduction of Frequency Division Multiplexing (FDM) techniques which required separate go and return channels. These techniques enabled more than one circuit to be carried over a single metallic or radio link. The 4kHz telephony signals modulated the amplitude of carriers which were spaced 4kHz apart (single sideband techniques being used) to form the basic 12-channel carrier group. These techniques became common place in the main network with a fixed hierarchy of channel spacings and multiplex structures being established to form supergroups and hypergroups. FDM techniques were also used in the junction network along with the introduction of induction coils and cable balancing to improve the reach and performance of these cables.

In the same way that the speech signal was carried in a 4kHz analogue form signalling was also constrained to the same frequency band and always constrained to the same channel. In the local network, loop disconnect signalling was and is still used in the majority of cases between the telephone and the local exchange. In strowger exchanges the equipment is operated directly by the breaks in the DC current and this same technique was used over the junction network by the local exchange to operate the main switching exchange. When a carrier system was encountered, this signal had to be translated from the near DC loop disconnect signal to a voice frequency signal within the 300 to 3400Hz channel band (generally 1800Hz).

In the early analogue network control was very basic with most exchanges being operated directly by the signalling. Early translation equipment in the form of the registers in the Director areas and the STD register translation equipment used relays and no further flexibility was possible.

12.3 DIGITAL TRANSMISSION

Digital transmission in the form of telegraphy of course pre-dates analogue telephony by a number of years but the carrying of the analogue telephony signal in digital form was not invented until the 1930s when Pulse Code Modulation techniques were first patented. However this technique together with time division multiplex techniques did not become commercially viable until the 1960s when they were first used in the junction network for routes of over 8 miles. These were 24 channel systems and by the end of the 60s the British Post Office had nearly 1000 systems installed. These 24 channel systems have now been superceded in Europe by a 30 channel system which operates at 2048kbit/s and conforms to CCITT recommendation G732. In this system the 4kHz analogue signal is sampled at 8kHz and the resulting samples quantised into 256 levels according to the A law logarithmic code and encoded into 8 bit bytes to produce a rate of 64kbit/s. Thirty of these channels are then assembled, along with a 64kbit/s signalling channel and a 64kbit/s framing channel, into a 2048kbit/s system.

Digital transmission has gradually become more cost effective to such an extent that a higher order TDM structure has been estabished at rates up to 565Mbit/s and even higher rates are being considered. These are now cheaper than the original FDM techniques both in the main and junction networks especially with the introduction of optical fibre. British Telecom therefore plans to convert its main transmission network, together with the director area tandem networks, to digital operation based on 64kbit/s channels by 1992. The resulting progressive penetration of digital transmissions systems is shown in Table 1.

TABLE 1. Penetration of Digital Transmission Systems

Type of circuit	1981-82 (%)	1986-87 (%)	1991-92 (%)
Analogue (Cable and microwave)	98	72	19
Digital (Cable, microwave and optical fibre)	2	28	81

12.4 DIGITAL EXCHANGES

The introduction of digital signalling and stored program control is inevitably influenced by the introduction of new exchanges. However, although the first forms of signalling were digital, the first modern form of digital signalling was used in the original 2Mbit/s PCM transmission systems. Each channel in the system is allocated four bits in time-slot 16 and these are used to convey the loop condition and other states. Specific signalling bits are therefore still associated directly with a particular channel and at the multiplexer they are translated back to an in-band signalling system for the exchange to handle.

Specialised Data Networks have seen the introduction of message based signalling systems, for example X25 in the Packet Switched Network, but this type of signalling was not introduced into the PSTN until the introduction of stored program control exchanges. Common channel message based signalling systems, where messages relating to different connections are statistically interleaved on a common channel, were then introduced to carry messages between exchanges. In the UK this occurred at the same time as the introduction of digital switches in the form of System X but in other countries they were used on analogue SPC exchanges. Loop-disconnect and Multi-frequency signalling continued to be used between the customer's terminal and the local exchange.

In the same way that technology had made the use of digital techniques in the transmission field more attractive, so digital exchanges became cheaper to implement and maintain. These digital exchanges are designed to handle 64kbit/s channels transparently using both space and time division digital switches. Main network digital switches became particularly attractive because of the existance of digital transmission and as a consequence the saving that can be made in analogue to digital conversion equipment. These digital main network exchanges together with the digital transmission form what is know as the Integrated Digital Network (IDN). The first System X exchanges were put into the main network in 1980 and it is hoped that by 1986 the thirty major cities in the UK will be linked by the IDN and that the whole of the main network will be digital by the early 1990s.

Digital local exchanges also became attractive, avoiding the analogue to digital conversion on the main network side of the exchange and these are currently being introduced into the network. It is planned that 650 sites will have a digital switching presence by the end of 1986, rising to some 1200 sites by the end of 1987. By the end of 1985, 1066 equipment orders for new sites and extensions had been placed covering some 3.2 million customer connections. BT's programme for the introduction of these exchanges is illustrated in the graph of Fig.12.1.

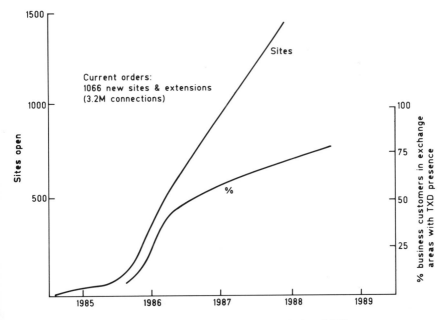

FIG. 12.1 PENETRATION OF DIGITAL LOCAL EXCHANGES

The conversion from the analogue signal produced by the standard telephone to a digital signal occupying 64kbit/s is now occurring at the front of the local exchange. Therefore connections across the main network between local exchanges are therefore able to provide:-

1) A completely transparent 64kbit/s channel.

2) A powerful digital signalling capability.

3) The flexibility provided by stored program control exchanges.

These facilities have been provided in order to support telephony in the most economical way possible. They have however set the scene for the development of what has become know as the ISDN, Integrated Services Digital Network.

12.5 INITIAL ISDN DEVELOPMENTS

A definition of the ISDN is as follows:

"A network that provides or supports a range of different telecommunication services via digital connections between user-network interfaces"

CCITT recommendations state that "ISDNs will be based on the concepts developed for telephony IDNs and may evolve by progressively incorporating additional functions and network features including those of any other dedicated networks such as circuit switching and packet switching for data so as to provide for existing and new services".

In common with most telecommunications administrations, BT wished to fully exploit the features of its emerging Integrated Digital Network (IDN) in order to extend the range of services offered to its customers. BT saw the early introduction of an ISDN Pilot Service as a means of gaining practical experience of developing and operating an ISDN and stimulating customer interest, and planning for this was started in 1979.

At this time it was decided to offer two new types of customer access and to market them under the name Integrated Digital Access (IDA). These were a single-line access and a multi-line access as shown in Fig.12.2 and were structured as follows:-

Single-line IDA - a 64kbit/s channel for speech or data
                  an 8kbit/s channel for data only
                  an 8kbit/s channel for signalling only

Total 80kbit/s plus additional framing information

Multi-line IDA - 30 x 64kbit/s channels for speech or data
            1 x 64kbit/s channel for signalling
            1 x 64kbit/s channel for framing and
              alarms

Total 2048kbit/s as for the PCM primary multiplex structure

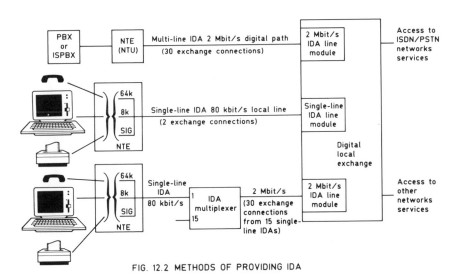

FIG. 12.2 METHODS OF PROVIDING IDA

In order to introduce these new services a number of new developments were required especially for the single-line IDA, these included:-

- Digital transmission in the local network with a bothway information rate of 80kbit/s between the customer's premises and the local exchange.

- Message based digital customer-network signalling.

- New types of customer's Network Terminating Equipment with standardised interfaces for the connection of data terminals

- New line exchange line interface units.

- A range of new call handling features such as closed user groups and called line identity.

These developments are described in more detail in the following sections.

12.6 LOCAL NETWORK TRANSMISSION

A cable network has already been established in the local area to provide telephony to customers and this represents a large proportion of British Telecoms capital assets. Although this cable has been installed to carry telephony with a bandwidth of less than 4kHz, it can be exploited to carry the wider bandwidth digital signals of the single line IDA link shown in Fig.12.2. The bandwidths for the single line access will be of the order of 100 to 200kHz depending upon the digital information rate and the modulation technique employed and as a consequence will be subjected to much greater attenuation (up to 60dB) than the telephony signal (a maximum of 10 or 15dB). Notwithstanding these high attenuations it is still feasible to provide full duplex transmission over the single telephony pair of wires by the adoption of suitable modulation techniques.

In its pilot service BT will be using two techniques for providing 80kbit/s transmission over the local network. The majority of customers will use a burst mode technique (Bird et al [1]) with a WAL2 line code, which ensures there is sufficient timing information present and that no equalisation of the characteristic distortion of the line is required at the receiver. The system is therefore relatively simple and therefore cheap to implement. This burst-mode system has a maximum loss of 34dB and will give a range of approximately 2.5km over 0.4mm copper pairs enabling 78% of customers to be connected directly to their local exchange.

A more complex echo cancelling technique (Adams et al [2]) is also being used on a smaller experimental scale, this system has a loss of 30dB and will provide a reach of approximately 3km enabling 89% of customers to be connected directly. More detailed information on these transmission techniques have been covered previously in Chapter 3.

12.7 SIGNALLING

System X digital exchanges are being introduced to provide cheaper switching for telephony customers and so support the standard analogue telephony interface. This meant that the customer's line module had to interpret the loop-disconnect signalling and pass the appropriate messages to the processor and also convert the analogue signal to the standard PCM digital format. The introduction of IDA required the development of a new digital subscribers line module. This module provides the appropriate digital transmission system and separates the 64kbit/s and 8kbit/s switched channels and the 8kbit/s digital signalling channel, directing them towards the switch and processor respectively. However no CCITT recommendations on digital access signalling protocols were available at the time of specification and so the definition of a totally new digital signalling system was required.

A new signalling system DASS (Digital Access Signalling System) was therefore defined (Bimpson et al [3]). The structure of DASS is based upon the levels of the ISO (International Standards Organisation) model for OSI (Open Systems Interconnection) (ISO [4]). The first three levels are defined and are:-

Level 1	Physical	–	8kbit/s and 64kbit/s signalling channels of the single and multi-line access.
Level 2	Link Access Protocol	–	based upon High Level Data Link Control (HDLC) procedures
Level 3	Network	–	a new defined set of call control messages

The HDLC format at level 2, previously discussed in Chapter 4, has the following fields.

The Flag acts as a unique delimiter and additional circuitry ensures that the flag pattern never occurs in the bits in the rest of the frame

The Address Field is used to indicate whether the message is a command or a response, a signalling or maintenance message and to which channel it refers. (ie the 64kbit/s or 8kbit/s channel).

The Control Field is used to control the interchange and acknowledgement of level 2 frames.

The Information Field contains the signalling or level 3 message.

The Cyclic Redundancy Check Bits (CRC) enable the receiver to determine whether any errors have been incurred during transmission. Complete error protection is obtained by repeatedly retransmitting the frames until a positive acknowledgement is received from the receiver. (This is referred to as a "compelled signalling system")

At level 3 a complete set of call control messages have been defined to enable the establishment and clearing of connections and supplementary services. These can vary in length from a 24 byte Initial Service Request for a Data call, a 17 byte clearing message, to a single byte Call Accepted message. A typical sequence of level 3 messages for the establishment of a Data call is shown in Fig.12.3.

The introduction of DASS for multi-line access to PABXs, the development of digital PABXs and the use of digital leased circuits to form digital private networks, led to a need for a digital inter-PABX signalling system. BT and a

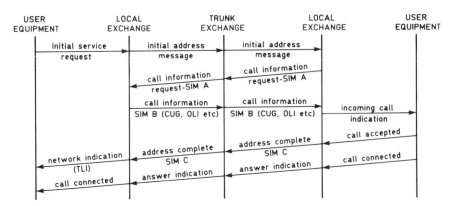

FIG. 12.3 MESSAGE SEQUENCE FOR BASIC ISDN CALL

number of UK PABX manufacturers therefore collaborated on the definition of a signalling system based upon DASS but which has been enhanced to meet the inter-PABX signalling requirement. This has been called the Digital Private Network Signalling System (DPNSS) [3]. During the definition of DPNSS it became apparent that it was desirable to align more closely certain of the messages at level 3 of DASS with those of DPNSS. At the same time proposals were being made to enhance the repetoire of messages at level 3 in order to provide more facilities to the user. An enhanced version of DASS No1 was therefore defined, called DASS No2, which would enable PABXs to have a common type of signalling system for both inter-PABX and PABX to Network Signalling. The common features of DPNSS and DASS enable them to be interleaved on a common signalling link such as time-slot 16 in a 2048kbit/s multiplex structure.

12.8 CALL HANDLING FEATURES

The introduction of a new message based signalling system into the local network enables a much wider range of facilities to be provided and for more information to be provided by the network on the progress of calls. However DASS No 1 as defined for the Pilot ISDN will initially only support the same supplementary services as defined for telephony customers and normally accessed by those customers using MF4 signalling (such as Call Diversion, Abbreviated Address, Three Party Call etc.) along with two new services for Data customers (Closed User Groups and Originating and Terminating Line Identities). This was in order to minimise the exchange software development required for the pilot ISDN service.

However, because of the low penetration of digital equipment in the early stages of IDA, users must now indicate at the beginning of the call the type of call they wish to make. There are three main types of ISDN calls, the ISDN telephony call, the data call and the voice/data call. Each of these types is identified to the network by means of the Service Indicator Codes (SIC).

12.8.1 ISDN Telephony Call

The ISDN customer with a telephone can make telephony calls to other ISDN customers who have telephones, or to any other customer connected to the world-wide telephone network. The calling customer will experience the usual tones and announcements that an ordinary System X telephony customer would hear.

12.8.2 ISDN Data Call

When a customer requires a data call, the Network Terminating Equipment (NTE) sends the appropiate SIC to the network. This tells the network that an all-digital transmission path is absolutely necessary and that common channel signalling must be employed throughout the routing.

For a data call, the routing digits from the NTE can only be sent in a single block. The network routes the call, checking as it goes for an all-digital routing. The network passes the same SIC to the called NTE that the first NTE sent to the network. This will tell the called NTE what type of call is being offered and hence the NTE can deduce which of its terminals would be able to accept the incoming call. If that terminal is free, then the terminal is alerted and the call can be answered. The path across the network is also protected from intrusion by telephone operators but in contrast to the ISDN Telephony call, only supervisory messages are sent to the customer and these are confined to the signalling channel.

In the rare situation where an analogue route is encountered on a data call set-up, the network clears down the call and sends back a message giving the reason why the call is being rejected.

12.8.3 ISDN Voice/Data Call

The calling customer indicates that a voice/data call is required. The network will set up the call as a voice connection but ensures, in a similar way to the straight data call, that an all-digital path is provided. If, during the call, the customers decide that they need to send a facsimile, a diagram for example, then they will be able to switch over to data and the appropiate terminals are connected to the line. The customers are then free to swap between the voice and data modes.

12.9 NETWORK TERMINATING EQUIPMENT

The principal functions provided by any single line IDA NTE are standard `X' and `V' series terminal interfaces, protocol conversion (interworking the signalling over the terminal interface with that used in the 8kbit/s signalling channel to the exchange), rate adaption as required (to bring the customers terminal information rate up to the 64kbit/s or 8kbit/s data rate of the traffic channels to be used), the multiplexing of the two traffic channels and the 8kbit/s signalling channel (total 80kbit/s) and the interface to the full duplex digital transmission system. In order to accommodate the wide range of terminals which might be connected, information relating to the particular terminals actually connected to each of the data ports is stored in the NTE, for example the information data rate of the terminal.

The rate adaption function of the NTE involves the mapping of the CCITT standard synchronous data rates (2.4, 4.8, 9.6 and 48kbit/s) and the recognised asynchronous rates presented on the data port to the 8 and 64kbit/s traffic channels. At the time of development no internationally recognised standard technique for adapting these rates to both the 8 and 64kbit/s rate was available. British Telecom therefore implemented its own techniques (based upon certain standards that did exist) which enabled all recognised synchronous rates up to 64kbit/s to be mapped onto the 64kbit/s bearer and rates up to 4.8kbit/s to be mapped onto the 8kbit/s bearer channel. Similarly asynchronous rates of up to 9.6kbit/s and 1.2kbit/s could be mapped onto the 64 and 8kbit/s bearers respectively.

Initially, for the pilot service, two NTE's were developed. The NTE1 was designed as a desk top instrument and includes a digital telephone, keypad, display and single data port. This data port may be configured with a CCITT X21 bis interface (or V24 via an external adaptor), or the leased circuit version of X21, to operate over the 8 or 64kbit/s channels to the exchange.

Call set-up is by means of the in-built keypad, and an alphanumeric display assists the user with call set-up and facility programming, ie the input of terminal-dependent data. On-hook dialling is provided together with a number of special function and service request keys (some programmable by the customer) to simplify use of supplementary services, eg closed user groups and short code dialling.

The NTE1 was designed during the early stages of the ISDN development programme and prior to the privatisation of BT.

As the NTE1 incorporates features of a terminal as well as those of an NTE, it does not meet the conditions for the liberalisation of customer equipment laid down by the government. Consequencially it will not form part of the generally available ISDN equipment; its use was restricted to the initial phase of the Pilot ISDN.

The NTE3 was designed for applications where there are a number of different terminal equipments requiring access to the BT network and where all are capable of controlling call set-up procedures. For this reason the NTE has no built-in telephone, keypad or display, but it has six terminal ports each of which may be configured to support one of the following terminals. An X21 interface (leased or circuit switched variant); an X21 bis interface (or V24 an external adaptor; a 2-wire analogue interface or a X24/V28 interface. The latter providing access via a modec (a modem and codec) to PSTN basic modems at rates up to 300bit/s. Because not more than two of the six terminal ports may be connected to the exchange at one time, the NTE resolves contention between them, acting effectively as a traffic concentrator.

New standard rate adaption techniques have now been agreed by CCITT for synchronous rates (CCITT recommendation X30). In addition the European Computer Manufacturers Association (ECMA) have agreed a technique for mapping the asynchronous rates up to 19.2 and 4.8kbit/s onto the 64 and 8kbit/s bearers respectively. British Telecom has now adopted these techniques in a new cost reduced NTE, NTE4, which was developed to replace the original NTEs and for wider use when the ISDN was extended in the middle of 1986.

The basic NTE4 will be a data only NTE offering two X21 ports supporting both circuit switched and leased circuit variants for synchronous data rates up to 64kbit/s. There will also be a V24 port for use only for facility programming. In order to facilitate interworking to V-series terminals and to provide a telephony capability a range of terminal adaptors were planned to connect onto the X21 port of the NTE4. The first adaptor to be developed was that providing an X21 bis interface (or V-24 via a passive adaptor) along with a digital telephone with an X21 interface.

12.10 THE PILOT NETWORK

In June of 1985 an ISDN Pilot Service was opened in the UK, based on a System X Local Exchange in the City of London, under the marketing name IDA (Integrated Digital Access). Early in 1986 it was extended to three further System X exchanges located in Birmingham, Manchester and London, which are fully interconnected via their respective digital main network switching centres with digital transmission links using CCITT No 7 (BT) signalling.

Although initially based only on these four Local Exchanges, the Pilot Service is extended to more than 60 additional exchange areas in the major centres of population by means of small remote multiplexers sited in analogue local exchanges.

These multiplexers house Digital Subscribers Line Units identical to those used by customers directly connected to a System X exchange, supporting the same basic IDA access. The multiplexer supports 15 customers' 80kbit/s links and multiplexers the 64kbit/s B channels into TSs1 to 15 of a standard 2Mbit/s structure (conforming to CCITT recommendation G732) and the 8kbit/s B' channels are reiterated up to 64kbit/s before occupying TSs17-31. The signalling messages contained within the customer's 8kbit/s signalling channel are then statistically interleaved in TS16. The 2Mbit/s output from the multiplexer then passes over standard digital line systems to the nearest System X local exchange where it terminates on the Digital Subscribers Switching Subsystem in the same way as the multi-line IDA. The remote multiplexer is also able to respond to maintenance messages from the exchange carried in TS16 which enables remote tests to be performed and to report alarms to the parent local exchange by sending messages in TS16.

These multiplexers together with the directly connected access circuits provide a total of 1000 basic access circuits. In the latter half of 1986 the Pilot Service will be further extended to an additional 130 or so sites with the introduction of more System X Local Exchanges and the new NTE4 as described earlier .

12.11 ISDN SERVICES

Now that the full 64kbit/s capability has been extended to the customer let us consider the resulting benefits that will be perceived by the customer when accessing both existing and new types of service.

12.11.1 Telephony

A complete digital transmission path will mean that voice quality transmission should be much better with little or no noise and interference from extraneous sources. 64kbit/s transmission should also give the opportunity for using different encoding techniques which provide a wider bandwidth, such as 7kHz speech transmission, alternatively other techniques could be used to carry more than one voice channel provided they are all destined for the same remote customer. Customer will also see the benefits of digital signalling with fast call set-up and a range of advanced supplementary services.

12.11.2 Prestel

The existing Prestel service operates at 1200bit/s and displays a page of information in a few seconds, increasing this rate to either 8 or 64kbit/s will result in a corresponding increase in page transmission times which is very useful when browsing through pages. However the advantage of ISDN to Picture Prestel, when a 3 x 3 inch picture would take over a minute to transmit, is more obvious, the initial picture being displayed in less than a second and the colour being completed in less than 5 seconds at 64kbit/s.

12.11.3 Facsimile

Current analogue facsimile techniques can take 3 or 6 minutes to transmitt an A4 size page, with 64kbit/s this rate is reduced to between 6 and 10 seconds and provides better quality, making facsimilie much more attractive to the user. Further new compression techniques, which take advantage of the better error performance, will enable this rate to reduced even further.

12.11.4 Slow-scan TV

The refresh time for a slow-scan TV frame on ISDN is 3 to 4 seconds. This speed of picture replenishment makes SSTV viable for remote security surveillence and for medical diagnostic purposes.

12.11.5 Teletex

This document transmission service will again benefit from the decreased transmission time, taking about half a second to send an A4 size page. This may not be too noticeable for short documents but would prove valuable for long file transfers.

12.11.6 Data

Although the above services are examples of structured data services, ISDN provides a completely transparent path over which any form of data can be transmitted. Large computer file transfers will thus benefit from the increased rate of transmission.

12.12 A NATIONAL ISDN

During the period in which BT was specifying and developing equipment for its ISDN Pilot Service work began in earnest within CCITT to define a set of recommendations relating to ISDN. This culminated in the I-series Recommendations approved at the 1984 CCITT Plenary Assembly (CCITT [5]). These recommendations differ significantly from the standards adopted by BT for the Pilot Service, particularly in terms of the access channel structure, signalling protocols, and rate adaptation.

During the same period BT was being transformed from a Public Corporation into a Private Company. The consequent introduction of competition in the supply of customers' terminal equipment resulted in the need for a clearly-defined customer-network interface as defined in the new I-series recommendations. Plans have therefore developed to move as soon as possible from the Pilot Service standards for basic access to new ones based on the CCITT recommendations.

However it was recognised that these I-series Recommendations covering basic access were by no means complete, and BT has been very active in progressing them further, both within CCITT and CEPT, in order to derive a complete and unambiguous set of specifications. These have now been defined in the form of a BT Network Requirement. Exchange developments require long development times, it was therefore decided to develop a new multiplexer, supporting 14 or 15 I-series basic access lines, but interfacing to the Local Exchange by means of the existing 2048kbit/s port with DASS signalling in TS16 as shown in Fig 12.4. This new multiplexer thus has to perform protocol conversion between the I-series signalling on the customer side and the DASS signalling on the exchange side. In mid-1985 specifications were prepared for the new multiplexer and in March 1986 a contract let for the supply of a number of these multiplexers for delivery before the end of 1987.

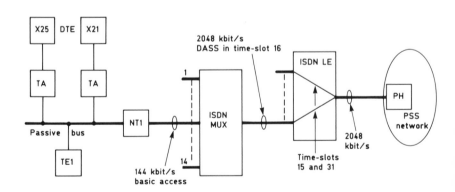

FIG. 12.4 PROVISION OF I-SERIES BASIC ACCESS

The need eventually to move from this DASS-based Local Exchange interface to one based on the I-series Recommendations for primary rate access at 2048kbit/s is fully recognised, and this is seen as the next step in the evolution of BT's ISDN. However, this is unlikely to occur until a firm and detailed recommendation is established meeting the requirements of modern PBXs for network and inter-PBX signalling. It is thought that this is unlikely to be achieved before the Plenary Assembly in 1988. Indeed, it has been clearly identified to PBX manufacturers in the UK that BT envisages the 2048kbit/s primary rate interface with DASS (and DPNSS) signalling enduring until well into the next decade.

12.13 THE I420 INTERFACE

In the CCITT recommendations two 64kbit/s B-channels are supported to the customer along with a 16kbit/s D-channel for signalling, giving a total information rate of 144kbit/s over the local network. The transmission system that supports this rate is contained within a Network Terminating equipment (NT1) at the customers' premises, see Fig.12.4. This equipment then supports the user/network interface to the customers' equipment, designated the I420 interface within the recommendations. Further recommendations give detailed descriptions of the physical layer (I430), link layer (I440/1) and network layer (I450/1) of this interface.

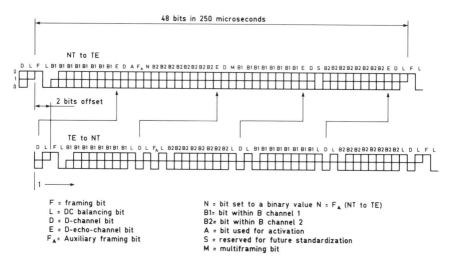

F = framing bit
L = DC balancing bit
D = D-channel bit
E = D-echo-channel bit
F$_A$ = Auxiliary framing bit

N = bit set to a binary value N = F$_A$ (NT to TE)
B1= bit within B channel 1
B2= bit within B channel 2
A = bit used for activation
S = reserved for future standardization
M = multiframing bit

FIG. 12.5 I420 FRAME STRUCTURE

The physical layer operates at 192kbit/s with a frame structure as illustrated in Fig.12.5. The rate is in excess of the 144kbit/s referred to earlier since the interface has been designed to support not only a point-to-point operation but also a point-to-multipoint operation when additional bits are required. In the latter case, when the interface acts as a passive bus, the D-channel has to be shared dynamically between up to 8 terminals. In order to resolve the contention for this channel a second D echo channel is provided from the NT1 to enable terminals to detect that they have successfully accessed the D-channel. Additional bits are also provided for Activation/Deactivation of the terminals, DC balancing of the transmission signal, since an AMI line code is employed, and for framing giving a total frame of 48 bits.

The link layer is known as the LAPD and like DASS is based on the HDLC procedures and formats described in section 12.7, but it is more closely aligned with the procedures of X25 LAPB. Layer 2 provides a data transfer service to the layer 3 functions of circuit switching, packet switching and maintenance/management. Within the 16kbit/s signalling channel logical connections called data links are established. Both point-to-point and point-to-multipoint (broadcast) data links exist in layer 2 between the terminals and the network. Each terminal has at least one point-point link along with access to the broadcast link. The procedures allow for unacknowledged data transfer on the broadcast link and acknowledged service on the point-to-point links. The acknowledged service includes a mechanism for recovery from frame loss due to line errors. Each data link is identified by the address field of the HDLC frame. The address consists of two parts; the service access point identifier (SAPI) identifying the layer 3 function and the terminal endpoint identifier (TEI) which identifies the individual data link. In addition to the basic data transfer and error recovery functions there are procedures defined for address allocation, link establishment, disconnection and flow control on the layer 2 link.

TABLE 12.2 Planned Supplementary Services on the National ISDN

Available on all Calls	Available on Telephony Calls
Closed User Group Call/Calling Line Indentity Call Charge Indication. Network Address Extension	Call Waiting Call Diversion Basic Diversion Diversion on Engaged Diversion on No Reply. Advice of Duration & charge Reminder Call. Incoming Call Barring. Outgoing Call Barring. Three Party Service

At layer 3 a format and set of basic call control messages have been established not unlike those defined for DASS, but further derfinition work is required to define more supplementary services and features. The interface that BT is introducing at the end of 1987 will support the supplementary services indentifiedin Table 12.2.

12.14 THE SUPPORT OF PACKET MODE TERMINALS IN AN ISDN

The demand for packet mode data services is increasing rapidly throughout the world as new developments in computer and information technology expose further opportunities to shape and serve the growing variety of user needs. There is no doubt that this trend will continue, even with the introduction of the fast circuit-switched data services which form an intrinsic part of an ISDN, since many of the new applications involve short interactive transactions for which packet switching is ideally suited. To meet this demand many countries have now established dedicated Packet Switched Public Data Networks (PSPDNs), in the UK this is BT's PSS network. Since these packet mode services are primarily aimed at the same users as the circuit-switched services directly supported by the emerging ISDN, that is the business community, integrated access to the PSPDN for ISDN-based packet terminals is seen as an important element of service integration.

The arrangement for packet access in the Pilot service (Gleen [6], Lisle [7]) corresponds to the minimum integration scenario outlined in CCITT Recommendation I462 [5]. It provides for packet access only via the B or B' channels to X25 ports on the PSPDN each of which is dedicated to a specific packet terminal. These ports may be accessed on either a leased-line basis or a dial-in (but not dial-out) basis in conjunction with a Closed User Group to prevent unauthorised use by other customers. This arrangement supports all X1 data rates by using the intrinsic rate adaptation capability built into the Pilot service.

The new CCITT I-series Recommendations however additionally provide for packet access via the D-channel, the packet or p-frames being interleaved with the signalling or s-frames relating to the 64kbit/s B-channels. Using the D channel for packet access has a number of important advantages:-

- unlike the B-channels the D-channel is non-blocking

- it can support a number of packet mode terminals simultaneously

- it leaves the B-channels free for terminals which cannot use the D-channel

- it can probably be achieved at very low marginal cost

Considering the low signalling rate required to handle B-channel traffic, even when supplementary services or user-to-user signalling are involved, the 16kbit/s D-channel in the Basic ISDN user-network interface should be capable of supporting most packet mode terminals up to 9.6kbit/s in terms of X1 user rates, including multiple terminal configurations. It is therefore proposed to introduce D-channel packet access via the I series multiplexer described earlier in 1988.

Packet mode terminals operating at 48kbit/s or 64kbit/s clearly have to use a B-channel for packet access, however those operating at lower rates could use either a B or the D-channel. To avoid complex problems of channel selection for virtual calls incoming to ISDN-based packet terminals, the customer will register at subscription time either for the use of the B-channel or the D-channel for packet access.

The basic intention with D-channel packet access is to provide the user with what appears to him to be dedicated circuits connecting each of his packet terminals to X25 ports on the PSS network. Thus, each such terminal will be identified by the PSS network in the usual way by a unique X121 Network User Address (NUA) allocated at subscription time. This NUA will be registered against the user's ISDN number which identifies the user at the user/network interface. Each D-channel packet terminal will also be given a pre-assigned Terminal Endpoint Identifier (TEI) at subscription time.

As outlined earlier, and illustrated in Fig.12.4, the policy for providing the user with I-series basic access is by means of small multiplexers each connected to the ISDN Local Exchange by a standard 2048kbit/s interface, and standard line transmission system if remote from the Local Exchange. D-channel packet access requires the capability in the Multiplexer to take the p-frames from the users' D-channels and map them on to 64kbit/s time-slots for transmission through the Local Exchange to the PSS network, and vice versa in the other direction of transmission. This will be done on a statistically multiplexed basis using one or two 64kbit/s time-slots depending on the aggregate level of D-channel packet traffic, time-slots 15 and 31 are reserved for this. A Packet Handler capability will be added to the PSS network to handle the resulting multiplexed packet streams.

12.15 CONCLUDING DISCUSSIONS

BT decided to establish an ISDN Pilot Service as a means of getting practical experience of developing and operating an ISDN and stimulating the interest of terminal manufacturers and customers. In the absence of firm recommendations this was based on standards developed by BT. Impressive progress within CCITT and CEPT has resulted in the early publication of firm and detailed specifications for the basic access interface. Prompted by the requirements of a new regulatory environment, a new phase of ISDN development is planned based on the introduction of the new standards. This centres on the use of small multiplexers to provide I-series basic access interfaces to customers whilst using the existing 2048kbit/s exchange interface with DASS signalling.

The general aim in relation to packet access is to exploit the commercially attractive features of the new I-series Recommendations, that is multiple terminal capability and low marginal cost, whilst keeping development to a minimum. The proposals outlined in section 12.14 represent the simplest approach to this within the constraints imposed by the existing PSPDN, and it is recognised that these developments are but a step towards full service integration.

It is inevitable that the customer's demand for higher bit rate services will grow and for the network to evolve to support these higher rate connections. Video conferencing and computer file transfers are already demanding rates between 384 and 2048kbit/s and even 8448kbit/s. Broadcast quality TV signals are now being distributed via interactive cable TV networks and although these are initially in analogue form these will eventually be provided digitally at rates of 34 or 70Mbit/s. Work has therefore started within CCITT on the definition of Broadband ISDNs and a number of broadband trials are being conducted throughout the world. A lot more work has therefore to be completed before we see a truely Integrated Services Digital Network incorporating narrowband and broadband circuit switched connections as well as packet mode connections.

REFERENCES

1. BIRD,J.R., BYLANSKI,P., TRITTON,J.A. "Aspects of Transmission and Distribution in the Local Network". Telecommunications Transmission – into the digital era. IEE Conference Publication No193

2. ADAMS,P.F., GLEN, P.J., WOOLHOUSE ,S.P. "Echo Cancellation Applied to WAL2 Digital Transmission in the Local Network" Telecommunications Transmission – into the digital era. IEE Conference Publication No193

3. BIMPSON,A.D., RUMSEY,D.C., HIETT,A.E. "Customers' Interfaces in the ISDN" British Telecommunications Engineering Journal, Vol 5, Part 1, April 1985

4. ISO TC97/SC16/537, "Data Processing – Open Systems Interconnection – Basic Reference Model"

5. CCITT Recommendations of the Series I, Integrated Services Digital Network (ISDN). Red Book, Vol III, Fascicle III.5, 1984.

6. GLEEN K E, "ISDN - Interworking with Other Networks", Networks '85, Proc. European Computer Communications Conference, London, June 1985.

7. LISLE P H and WEDLAKE J O, "Data Services and the ISDN", British Telecommunications Engineering, Vol 3, July 1984.

Chapter 13

Cable television

Ken Quinton

13.1   INTRODUCTION

Cable television networks originated as broadcast relay systems and the majority of such systems worldwide have presented to the domestic TV receiving equipment several programmes at one coaxial connection arranged in frequency division multiplex, i.e. as occurs on the downlead from a TV aerial.   Unlike a telecommunications system, the TV signal is not transmitted at baseband.   The picture signal has a spectrum from 25 Hz to above 5 MHz, i.e. in excess of 17 octaves, which would cause extreme equalization problems and make inexpensive transformers impractical.

Many broadcast relay networks were commissioned in the 1950s and 60s in the UK which operated at HF and used the principle of space division multiplex, one programme per pair in a multipair cable.   As the number of programmes to be carried increased, this principle was abandoned.

During the last 4 to 5 years, new targets have been provided for system designers.   Firstly, provision to be made for incoming signals from both direct broadcast and communications geostationary satellites.   Secondly, there is an increasing interest in local community activities, including entertainment, information services, local advertising; also there are educational possibilities.   Thirdly, with a suitably engineered system, there are possibilities for various levels of interactive communications, almost all of which are best served by a network which is configured as a 2-way point-to-multipoint communication system.   It will be seen that this communication system requires a much greater bandwidth in the "downstream" direction than in the "upstream" direction;   information in the latter will normally be carried in a digital form.

The idea of exploiting the potential of wide band communication systems is not very new;   a report by a working party of the National Electronics Council produced in October 1973 outlined a study programme.(1)

An essential element in these new developments is the control of access by the cable operator to at least some of the non-broadcast TV signals, either because they will be available on a subscription basis, or a pay-per-view basis, or because some channels could be used for a private audience and/or for business purposes, etc.

Many of the potential new services were outlined in an ITAP Report "Cable Systems".(2)

Evolving technology has its part to play and as cable systems develop increasing use is being made of real-time computing, and optical fibres as a transmission medium are expected to become commonplace within the next few years.

13.2   OVERALL TECHNICAL REQUIREMENTS

The IEC Specification N° 728 and the British Standard N° 6513 give minimum performance requirements for cable systems delivering TV and Band II FM to subscribers from a central point.   In compiling the specifications, the many sources of signal degradation were taken into account. These include signal levels and signal/noise ratios, inter-modulation and crossmodulation arising from equipment non-linearity, echoes which arise from mismatched junctions in the distribution path (which affect picture and teletext signals in different ways), the linear distortions of ampli-tude and group delay versus frequency, etc.   Methods of measurement are also detailed.

Additionally, BS 6513 deals with the safety aspects, in particular with regard to safety in the home.

There are Government-imposed constraints on the per-mitted maximum radiation from cable systems, and these are covered in the documents MPT 1510 and MPT 1520 issued by the Radio Regulatory Department of the Department of Trade and Industry.   Clearly, it is also in the interests of the cable system operator that the network should be adequately screened against ingress of interference and it should be noted that, whilst good screening is required where signal levels are high, to control radiation, screening to prevent interference will be concentrated where the signal levels are low.

13.2.1 DBS Signals

The current IEC and BSI specifications presuppose the carriage of TV signals to existing broadcast standards. Sig-nals from direct broadcast satellites expected to be giving public service to some European countries during 1987 will be to a new TV standard and it is anticipated that before 1990 there will be a UK DBS service also using the new MAC standard.   In this, the sound, chrominance and luminance sig-nals are carried in time division multiplex whereas the present terrestrial standards use frequency division multi-plex.   The MAC signals as expected to be carried on cable using amplitude modulation (the broadcast through the satellite transponder uses frequency modulation) have been described in an EBU publication, SPB 352.   The carriage of these MAC signals will be more demanding in terms of band-width per channel and permitted echo amplitudes because the picture signal is time compressed during transmission and the sound and data signals are digital either at 20 Mbs/sec or 10 Mbs/sec, using duo-binary coding.   However, the MAC signal is expected to be less demanding in terms of amplifier linearity because the subcarriers used for transmission of chrominance and sound have been avoided.

13.3 PRIMARY DISTRIBUTION NETWORKS

13.3.1 Super Trunks

The purpose of a super trunk is to carry a multiplicity of TV signals from a source to a nodal point from which a true distribution network originates. It could, for example, connect from a city centre to the centre of a large suburb, and, because almost all of the permitted signal degradation will be allocated to the installation beyond the node, the performance of the super trunk has to be close to perfection. There are 3 options :

(a) Microwave linkage, which might operate to CCITT standards, but when a large number of TV signals is to be carried it is more usual to impress upon the frequency modulating element for the klystron at the sending terminal the several TV signals arranged in FDM as would be applied to a cable.

(b) Use of a low-loss coaxial cable with each TV signal carried by frequency modulation, choosing a deviation ratio as large as possible (to obtain good s/n performance) consistent with the number of channels to be carried within the equalized spectrum.

(c) By using optical fibres. If frequency modulated signals are applied to the intensity modulator then, with a monomode fibre and light wavelength of 1300 nm, a distance of about 30 km can currently be covered with 4 TV signals/fibre. Alternatively, the baseband TV signal can be digitally encoded using PCM and, by using regenerators at repeater points, satisfactory transmission becomes possible over very great distances.

13.3.2 The Primary Distribution Network

Normally, this will comprise low-loss coaxial cables carrying TV signals in the downstream direction, the lowest carrier frequency being around 50 MHz and the highest between 150 and 450 MHz. Below 50 MHz will be provision for one upstream TV channel and 2-way data. The ability to carry a TV signal upstream is a licence requirement for new networks in both the UK and the USA. In order to minimize intermodulation effects, it is common practice to arrange the downstream vision carriers to be exactly equally spaced in frequency, a technique known as harmonically related carriers (HRC).

Most of the technical design details have been described in a paper by Paul Villé.(3)

Generally, the primary distribution network will be in a branching configuration leading to amplifier stations or switching points beyond which connections are made to subcribers' premises.

The primary distribution network adopted by British Telecom (4) has optical fibres connected in a branching configuration. Each carries 4 TV signals each of which is in a frequency modulated form when applied to the intensity modulator.

For economic reasons, the primary distribution network is permitted to contribute 25% or less of the overall permitted signal degradation because most of the cable and apparatus will be in the secondary distribution.

13.4   RECENT DESIGN OBJECTIVES

Annex B of the ITAP Report (2) is reproduced in Figure 13.1. A cable system to meet these broad requirements will be substantially different to one devoted solely to broadcast relay. Further, a new community-wide network cannot now be installed in the UK without (a) a licence from the Cable Authority and (b) an associated technical licence. These licences will cause all cable to be positioned underground unless there is a convenient existing pole and even then the pole can only support the final drop cable into a dwelling. Thus the initial installation must be capable of new and many services and the licence will terminate at 15 years unless, by Year 6, switching points are installed to permit interactive operations, whereupon the licence is for 23 years.

Fig. 13.1 <u>Services potentially available on cable systems</u>

- Terrestrial TV and radio channels (from UK or overseas)
- Satellite TV and radio broadcasts (also from UK or overseas)
- Subscription TV (films, sport, arts, etc.)
- Specialised subject channels,
  e.g. news, education, religious programmes, health
- Specialised audience channels,
  e.g. for particular ethnic groups;
        for different age groups - children, aged, etc.;
        for people with impaired hearing
- Local channels,
  e.g. local and national government information;
        "What's On"; consumer information;
        programmes by and about community groups
- Other services

        fire & burglar alarms; control of heating systems;
        remote meter reading; shopping, banking, betting,
        etc., from home; opinion polling; video games;
        electronic mail/messaging; interactive computer-
        assisted learning; general videotex information;
        software supply to home computers; access to
        national and international communications;
        wideband business communications

Fig. 13.2 The dependence of cable services on technical provisions	Notes
1. ENTERTAINMENT	
1.1 Basic Tier	
5 Terrestrial broadcasts, BBC and ITV	
4 Cable-originated programmes	
1 or more non-premium satellite programmes	
1.2 Premium Programmes	
1 Cable-originated subscription TV	a
1 Sport	a b d
1 Arts	a c
1 or more premium satellite programmes	a
2. NON-ENTERTAINMENT, TV BASED	
2.1 Basic	
Programme Guide	
Cabletext	
Educational	b d
Local/National Government	e
2.2 Premium	
News	a
Financial News	a
2.3 Leased	
2.3.1 Basic	
Shopping, Banking, Insurance,	b
Building Societies, Betting, etc.	b
Private broadcasting	a
2.3.2 Premium	
Home Study Courses	d
3. NON-ENTERTAINMENT, DATA BASED (Leased or Premium)	
Security, Meter Reading, Audience Measurement, Funds Transfer, Downloaded Computer Software, Electronic Mail	)) f)

Notes : Associated Technical Provisions

(a)   Control of access to TV channels :

       Premium services and subscription TV
       Private broadcasting

(b)   Upstream communication by numeric keypad

(c)   Pay-per-view access control and billing

(d)   Upstream communication by alpha-numeric keypad

(e)   Polling

(f)   Data transmission :   Upstream
                       2-way

It must be appreciated that, apart from a few strictly limited experiments, cable systems in the UK and, indeed, in Europe except for an operation in Helsinki, were constrained to the provision of broadcast relay prior to this decade. Thus, the majority of the services mentioned in Figures 13.1 and 13.2 are novel to the medium and the popularity which they are likely to enjoy is largely unknown.   Some experience with interactive services has been gained in the USA, and a paper by P. J. Alden (5) provides a summary account and draws the conclusion that the typical American cable network has technical shortcomings which make some trials almost impossible.

13.4.1   Subscription TV

This has proved very successful in the USA and elsewhere; it is a means of supporting entertainment programme provision without resorting to a licence fee or advertising.   Referring to Figure 13.2, there will normally be a basic tier of services available to a cable TV subscriber for a minimum regular payment.   The subscription services are made available for additional regular payments, and it is desirable that any combination of these can be chosen.   The cable system therefore has to provide "conditional access" to subscription channels.   It is highly desirable that the control of conditional access should be by electronic command from the cable system operator rather than by any change of hardware because individual subscribers are likely to vary their requirements from time to time.

The reader should notice that subscription TV is referred to as pay-TV in North America.

From the operator's viewpoint, it is advantageous if access to the basic tier services can also be controlled electronically, to provide for family removals, bad debt situations, etc.

13.4.2   Pay-per-view

An "impulse buy" has been shown to be more profitable than a subscription arrangement for some programmes.   These might cover an international sports event or a live visit to the opera, ballet or theatre.

13.4.3   Private Broadcasting

Once electronic control of access is provided, the possibility exists of addressing a specialized audience.   This could comprise medical practitioners (for whom there might be an updating service or programmes from pharmaceutical companies), the police, fire services, etc.

13.4.4   Messaging via a Numeric Keypad

Inevitably, when a large number of TV programmes is made available, a numeric keypad will be provided for programme selection.   The same keypad can be used for responding from the home, thereby enabling a wide range of interactive services. Referring to Figure 13.2, this activity is associated with sport, for example, for placing a fixed-amount bet against a list of runners, for simple responses in connection with education, for requesting more information on an advertised product and for transactional services. In the latter case, the presentation from the service provider needs to suit numeric responses.

Normally, there will be a symbol on the keypad which ensures that the following numerals are treated as a message for routing to the operator (or gateway) and a second symbol which will indicate termination of the message so that following operations will initiate channel change.

A typical message to the cable system operator would be a request for a pay-per-view offering.

13.4.5 Message Peaks

It should be noted that messages from subscribers are likely to have large peaks at particular times and the communications arrangements and computer facilities must be so designed to deal with these. For example, the number of messages will increase rapidly immediately before a pay-per-view offering or an event involving betting starts whereas, if an advertiser offers a particular bargain "Order Now Whilst Stocks Last", there will also be a peak but this will decline rapidly after the event.

13.4.6 Alpha-numeric Keypads

Provision of an alpha-numeric keypad, probably as an accessory, permits a further dimension in interactive operations. The maximum data rate from the subscriber is still below 100 Baud and there are clear advantages in responding to educational programmes, in transaction activities, placing complex bets, etc. Additionally, this keypad is appropriate for access to Prestel, for originating electronic mail and other data communications. Readers might associate some of these activities with telecommunications rather than cable television networks but an advantage of the latter is the avoidance of a modem and of temporarily disabling the telephone service. The incoming message will be routed to the normal TV screen either by teletext or by conversion to an analogue TV signal within the cable TV network.

13.4.7 Data Communications

Item 3 on Figure 13.2 shows a variety of services which are not TV-related but for which a point-multipoint network has advantages over the point-point telecommunications system. For example, in a high security alarm service, it is necessary to poll all terminals frequently. Likewise, if a gas or electricity supplier wished to adjust his charges according to the time of day or control the peak load, then this would be facilitated if the network were connected to the meter points.

Considering down-loaded computer software, a typical data rate for the input to a PC is currently 1200 Baud. As technology advances, the data rate is likely to increase, and this should present no problems to a cable system designed for very wideband downstream delivery. Finally, it is a simple matter to arrange for data to pass to the cable operator giving the total audience for each programme, either continuously or on demand, and thus provide him with a valuable research input.

13.4.8   Sources of Revenue

A feature of the new non-entertainment services is that they add to the cable operator's sources of revenue. The programme guide, local community activities and information services by cabletext serve to augment the amenity value and create interest, thereby attracting more subscribers. Secondly, for many of the services, the cable operator will be acting as a carrier and his revenue will come from the service providers, public utilities, results of audience measurement, etc.   These additional incomes are expected to justify the investment required for technical provisions beyond broadcast relay and subscription TV.

13.5   TREE AND BRANCH NETWORKS

All cable TV systems designed solely for broadcast relay operations follow a tree and branch topology. Normally, the interface between the primary distribution and the secondary network is at an amplifier station where there is a secondary output terminal. This amplifier is usually termed a "trunk/ bridger" and amplifiers in the continuing secondary distribution are termed "line extenders".   The secondary network is routed down the streets and at frequent intervals, every 2 or 4 dwellings, passive taps are connected to feed drop cables terminating within the dwellings.

The spectrum occupancy is as in the primary network, see Section 13.3.2. So far as the subscriber is concerned, the TV programmes are presented in frequency division multiplex and he suits his requirements by tuning the receiver.

The layout of the secondary network is illustrated in Figure 13.3.

In the UK, a VHF-UHF frequency changer is required to precede the TV receiver, which is normally only available capable of UHF reception. A mixer is essentially a non-linear device and the number of programmes which can be applied without causing disturbing products is limited. For a large number of channels, the converter has to be selective, i.e. it will contain the tuner. The receiver will then be fixed tuned.

13.5.1 Subscription TV

To provide conditional access to one or more channels, various methods have been tried, such as band stop filters or the addition of an interfering CW with notch filters to attenuate this. These applications were very inflexible and the principle of scrambling the premium services is now commonplace. Descrambling is normally done at baseband, so the complete domestic unit comprises a tuner, demodulator, descrambler and output modulator. Several programmes might be scrambled, more than those to which the subscriber is entitled, so the unit is made addressable by data signal from the cable head-end to effect the required operations.

13.5.2 Interactive Services

These are associated with data communications and the tree and branch network has limitations :

(a) The spectrum available is restricted and the only nodal points where data could be processed to economize in primary network bandwidth usage are the trunk/bridger locations and suitable processor items have not been developed. Current practice is to allocate data carriers to subscribers which will be disposed in FDM.

(b) There is a lack of security; i.e. one subscriber has access to the communications to and from another.

(c) There is a signal/noise problem on upstream communications. Whereas, downstream to a subscriber, there might be, say, 25 amplifiers in tandem, all of the upstream amplifier stations throughout the network will contribute noise at the cable head-end.

The usual way of dealing with this problem is to provide switches at the trunk/bridger locations and these are polled, only a few through connections being made at any one time. This practice leads to considerable delay in upstream messaging. The reader is referred to a paper by G. Allora-Abbondi. (6)

The use of analogue techniques for upstream data, to conserve bandwidth, results in the signals arriving at the head-end being prone to interference.

13.5.3 Fibre Optics

At the present time, a final distribution network in branching optical fibres is not a practical proposition. Optical tapping devices are neither as cheap nor as efficient as their electrical equivalents, and the permissible attenuation between inexpensive sending and receiving devices would allow very few splitters to be connected in tandem. With the present state of the art, it would require several fibres to carry 30 TV programmes.

13.5.4 Advantages

The main advantage of tree and branch secondary network construction is that the investment prior to connection of the first subscriber is comparatively low. Even when each subscriber has one addressable set-top unit, delivering UHF,

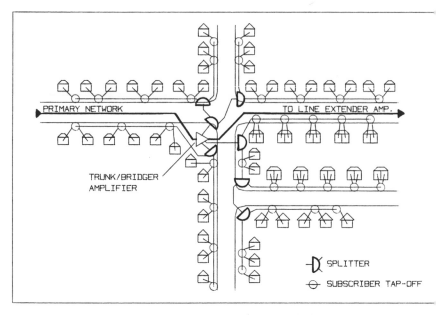

Fig. 13.3   Branching secondary network

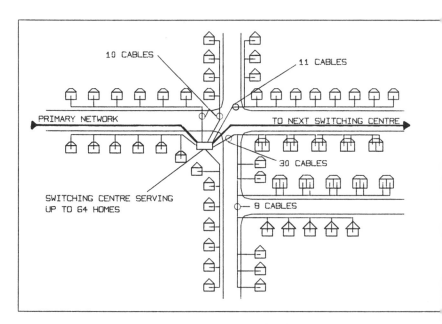

Fig. 13.4   Star secondary network

he total cost is still favourable. Available evidence
uggests that the tree and branch system is disadvantaged
hen subscribers require 2 set-top units, to feed indepen-
ently either 2 receivers or a VCR and receiver.

3.6 MINI-HUB SYSTEMS

These systems were introduced in Holland and are charac-
erized by having nodal points where the primary network
nterfaces with a secondary distribution which is in star
ormat, i.e. individual cables to dwellings. They were
ntroduced for 2 reasons :

a) A high proportion of receivers have inadequate selecti-
vity to receive adjacent channels. The solution was
to transmit on adjacent VHF channels in the primary
network but to convert at the mini-hub so that delivery
to subscribers was on alternate channels, some being in
the UHF spectrum.

b) To provide a layout which would permit later conversion
to switched star operation.

he philosophy is outlined in a paper by Dr. A. P. Bolle.(7)

The mini-hub is more expensive than a tree and branch
nstallation.

3.7 SWITCHED STAR NETWORKS

In a switched star system, the primary network is
sually in a branching configuration and it interfaces with
he secondary network at switching points beyond which the
istribution is by a discrete cable to each dwelling. This
dea originated from :

a) A rural requirement for a long drop cable which, for
cheapness, should only carry the 1 or 2 TV programmes
being viewed by the family. (8)(9)

b) The concept of an ever increasing number of TV programmes, which could be dealt with by augmenting the
primary network and the switch whilst the secondary
distribution (which comprises a high proportion of the
total cable required) could remain undisturbed.(10)

There are 3 principles on which the switching centre
an be based. Firstly, by providing a total of crosspoints
hich is the product of the number of incoming channels times
he number of independent outlets. This was used in the
SA and Holland and is the basis of the BT switched system
nstalled in Westminster.(4) Secondly, the "switch" can be
 frequency-agile converter, i.e. effectively the receiver
uner positioned in the switching point but with an output
requency suitable for the star network. In this case
here is one converter per independent outlet. This prin-
iple is in use in the UK by GEC and Cabletime. Thirdly, a
ybrid arrangement of crosspoint switching between a plura-
ity of incoming cables with each output from the crosspoint

matrix connected to a frequency-agile converter.   This
principle has been in use by BCS for several years.

Each of the above arrangements has its merits but, in
the first case, there can be a lack of transparency if the
switching is done at baseband because a demodulation/remodu-
lation process is involved.

The layout of the secondary network is illustrated in
Figure 13.4.

13.7.1   Conditional Access

The control of a switch by a subscriber can be over-
ridden by cable operator control to effect programme denial.
This is important because scrambling, with its attendant
possibilities for signal degradation, is avoided.   It is
doubly important if signal waveforms are subject to evolution
or the channels are likely to be used for other purposes,
e.g. very high speed data, because the many descramblers
will then have been tailored to the original waveform.

Channel denial information can be downloaded and stored
in the switching points, which is more economic than storage
in domestic units.

13.7.2   Pay-per-view

There is no need for the orders to reach the head-end
before a programme can be released.   The switching point
memory can be informed periodically regarding the credit-
-worthy subscribers and, when a PPV service is taken, the
head-end billing system can be advised at a suitable time.

13.7.3   Interactive Services

It was shown in Sections 13.4.6 and 13.4.7 that these
involve upstream data communications and, in some cases,
2-way data comms.   In a switched star system, the problems
associated with tree and branch networks are overcome by
using polling techniques between the switching points and
the subscribers and adopting data packet principles between
switching points and the cable system head-end.

13.7.4   Data Communications

Currently, switching points are designed for 16 - 300
subscribers so these can be polled several times per second,
thus 2-way low speed data communications require very little
storage in the domestic terminal of the cable system.

High speed data between subscribers and switching
points (where MUX/DEMUX processing occurs) can be allocated
a bandwidth of several MHz, if required.

13.7.5   Subscriber Responses

Messages for subscribers' TV screens can be originated
in switching points, e.g. by arranging for downstream Prestel
signals to be converted in equipment shared by a few subscri-

ers.   Using such analogue picture generators, it is also
ossible to originate to individual homes a range of messages
eld in store such as "Thank you for your order", "Your VCR
s receiving Channel X", etc.   These communications are considered to be essential for many of the new services.

3.7.6   Domestic Equipment

Basically, this comprises a keypad for channel changing
and message sending, probably mobile with infra-red linkage,
and some simple data comms arrangement, with memory.   If
the incoming TV signals are at VHF, and there might be 2 or
3 in FDM, then a simple VHF-UHF converter is added.   Independent programme provision to receiver and VCR requires no
additional equipment (although an extra switch is required
in the switching point).
The low cost of the domestic items is an important consideration because their average life is comparatively short.

3.7.7   Security

Not only is it impossible for a neighbour to eavesdrop
on a subscriber's communications, because of the discrete
connection to the switching point, but theft of service on a
switched star system is also prevented.   With the tree and
branch and mini-hub systems, where all distributed signals
enter the home, there is always the possibility of equipment
being developed and obtained by the householder which will
give unauthorized access to services.

3.7.8   Advantages

Summarizing, switched star systems have these advantages:

- The number of TV programmes available to subscribers can
  be increased indefinitely without disturbance to the
  secondary network.

- Conditional access is achieved without expensive, usually
  addressable, equipment in the home, and it is impossible
  for the subscriber to gain unauthorized access.   The
  alternative of scrambling, with its attendant drawbacks,
  loss of signal fidelity and incapacity to accept a change
  of TV standards, is avoided.

- The delay on pay-per-view access is minimal.

- A large effective bandwidth for upstream communications is
  available, due to data processing in switching points.
  The message queuing which occurs with tree and branch systems due to node switching is avoided.

- Text signals can be inserted in switching points.

- Neighbours cannot eavesdrop on private communications.

- The topology lends itself to adoption of optical fibre
  transmission.

13.8   TRANSMISSION MEDIA

Currently, almost all new cable systems in the UK are using coaxial cables to BS 5425.   In Westminster, BT are using an optical fibre primary network in branching configuration and each fibre carries 4 TV signals.   Additionally, there is a star-configured overlaid primary network, provided for video library services.   In this overlay network, an advantage of fibre is exploited;   the attenuation is low enough that the furthest switching point is reached without intervening active equipment.

Considering secondary distribution, it is not economic at the present time to provide 2 pairs of optical semiconductors, for downstream and upstream operation, per subscriber. With time, this situation is likely to change and switched star systems using exclusively optical fibres will become a reality.

The likelihood of a true optical switch becoming practicable, and controllable electronically rather than mechanically, is a matter for conjecture.

REFERENCES

1.   Report on cable communication systems by National Electronics Council, 1974, National Electronics Review, 10/1, 12-14

2.   Report by Cabinet Office Information Technology Advisory Panel, 1982, "Cable Systems", HMSO, ISBN 0 11 6308214 *

3.   Villé, P., 1983, "Broadband Systems", Local Telecommunications, IEE Telecommunications Series 10 (First Edition), 171-192

4.   Ritchie, W. K., 1984, "The British Telecom switched-star cable TV network", Br Telecom Technol J, 2/4, 5-17

5.   Alden, P. J., 1985, "The American experience with interactive services", Record of the International TV Symposium, Montreux, 537-546

6.   Allora-Abbondi, G., 1986, "Interactive communications networks", Cable Television Engineering, 13/4, 157-164

7.   Bolle, Dr. A. P., 1976, "Three-network philosophy for local networks and future integration in local network", IEE Conference Publication 137, 190-192

8.   UK Patent 1.158.918, October 1966

9.   Tough, G. A. & Coyne, J. J., "Elie - an integrated broadband communication system using fiber optics", 1983, Proceedings of the National Cable Television Association Conference (USA), 202-206

10.   Gabriel, R. P., 1970, "Dial-a-Program - an HF remote selection cable television system", Proc. IEEE, 58/7, 1016-1023

* Annex B reproduced with permission of Controller of HMSO.

Chapter 14

Electronics in the local network

Granville Taylor

14.1 INTRODUCTION

Until recently, electronic systems have been used mainly to provide or improve basic telephone services over twisted pairs of wires. In the last few years, high speed data transmission has resulted in a significant increase in the amount of electronic equipment in the local network, and this trend will increase with the introduction of ISDN. Optical systems are also beginning to find their way into the local network, and the often anticipated day when the PSTN will be optically based certainly looks a little nearer today, but it will still be many years before pair based networks are superseded.

The present day application of electronics in the local network can be divided into four main areas:

1   Extension of signalling and transmission limits
2   Pair-gain systems
3   New services
4   Digital transmission

This chapter outlines the various systems involved, their limitations and principle design criteria. It is not concerned with customer's apparatus used to provide additional services on the PSTN eg data modems.

14.2 LINE EXTENDERS

Two factors constrain the maximum length of local lines - audio loss (measured at 1600 Hz) and loop resistance. The local-line audio loss forms part of a country's national transmission plan, and therefore varies from country to country, but within the limits set for international connexions by CCITT. In the UK, the limit is 10 dB at 1600 Hz. The maximum loop-resistance is dependent on the design of the local exchange, being determined by the minimum current required for exchange signalling and for correct operation of the telephone microphone. The UK limit is 1000-1500 ohms.

The limiting values of audio loss and loop resistance may be increased by inserting into the line at the exchange an audio amplifier and current booster respectively. Such devices are available separately, as the characteristics of the various cable gauges that make up the local network determine that one or other of the limits is the constraining factor. These Audio Line Extenders, and Signalling Line Extenders find application in four areas:

1   Improvement of transmission performance of long lines.
2   As part of a fine gauge planning policy where cheaper, small

gauge cables are installed in the local network and line extender used to compensate for the increased loss and loop resistance.
3   Where a small exchange is amalgamated into a larger one and the loss of the longer local cable is offset by the addition o line extenders.
4   The provision of low loss lines for PBXs (Premium PBX lin service).

14.2.1   Audio Line Extenders

These devices permit the maximum attenuation of the local line t(be increased to 15 dB at 1600 Hz.   Although usually provided on a pe) line basis, they are sometimes fitted within the switching equipmen' (eg: between the A and B switches in a TXE2) so that they can b(shared between a number of customers.   Two basic types of audi(line-extender are available:

a   Voice-switched   This type of amplifier gives 6 dB of gain a(1600 Hz in one direction and 8 dB loss in the other, the gair being switched in favour of the higher signal-level.   It i(virtually undetectable in use because of the limited change o) gain, and relatively fast switching transition.   The voice switchec line-extender has the great advantage of being stable under any terminating condition, as the loop gain is less than unity. Disadvantages are its unsuitability for carrying duplex data, and possible interaction with other voice-switched devices such as loudspeaking telephones, echo cancellers and other voice-switchec line-extenders in the connection.
b   Hybrid   Two hybrid transformers separate the two directions of transmission, with four-wire amplification between the hybrids (Fig 14.1).   The line-extenders give a gain of between 0 and 7 dB at 1600 Hz in each direction, variable in $\frac{1}{2}$ dB steps, whenever current flow is detected.   Separate balance networks are required to match into the local line and into the exchange. However, because the exchange impedance varies widely from call to call and during call set up, stability can only be guaranteed if

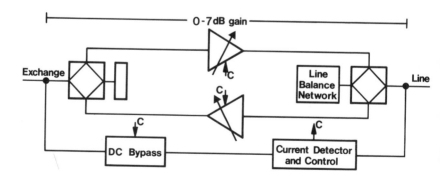

Fig 14.1   Hybrid Audio Line Extender

the return loss of the line impedance against the hybrid balance impedance exceeds 12 dB. This critical balancing has to be achieved against a wide variety of cable impedances and when the cable is terminated by open and short circuits during dialling. In practice, just one line balance impedance network has served the vast majority of situations, other balances being used for certain critical lengths of light and heavy gauge cable.

Both types of audio line-extenders are in use in the UK, the majority being the hybrid variety.

In the hybrid audio line-extender, separate DC bypass chokes rather than the hybrid transformers themselves are used for the transmission of signalling pulses and the DC loop current. In this way, the design of the hybrid transformers is eased and the resistance added to the local line is reduced. One system uses a flux-cancelling device to minimise the size of the bypass chokes; the core flux is detected by a hall-effect device and a DC current flowing in another winding on the choke is adjusted to cancel out the steady state flux due to the telephone loop current.

Further requirements are that the DC blocking capacitors at the centre of the hybrid must be disconnected when the line is idle in order that the capacitor in the telephone bell circuit may still be detected by the standard testing techniques; and special provision must be made in the voice-switched audio line extender for preferential switching in favour of multi-frequency signalling in the presence of dial tone.

14.2.2   Signalling Line Extenders

The usual method of extending the signalling limit of local lines beyond that normally permitted by the exchange equipment is to create a 'floating battery'. This may be connected in series with the line to boost the current flowing in the cable pair such that it does not fall below the design limits for the exchange equipment or below the acceptable microphone feeding current.

A 'floating battery' is obtained from the exchange battery using a DC/DC converter working at around 25 kHz; the oscillator frequency must be chosen to give good conversion efficiency, and to minimise the risk of interference with other equipment. The full boost voltage may

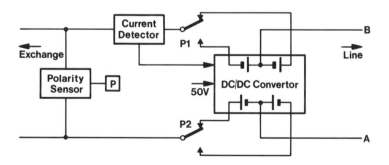

Fig 14.2   Signalling Line Extender

be added in either the A or B wire of the pair, or half in each (full battery boost, or split battery boost).

British Telecom uses signalling line-extenders with split battery boost of up to an additional 25 V in each wire of the pair, which ensures satisfactory signalling on lines up to 2000 ohms. Split battery boost is preferred as the balance of the pair is maintained. A block diagram of the system is shown in Fig 14.2. When current flow is detected, the DC/DC converter is energised, and the additional supply connected in the correct sense in each wire of the pair. This must be achieved quickly, as the system is required to be able to follow reversals in order to be compatible with various types of customer terminals. It must also cater for additional signalling conditions, such as earth loop recall on certain designs of PABX's, and 5 kohm loops on pay-on-answer coinbox lines.

The voltage added into the circuit must be carefully controlled so that on short lines, the regulator in 700 series telephones is still operative, and excessive current cannot pass through the exchange equipment. These short lines occur when signalling line extenders are provided to serve PABX's, or certain types of telephone installations, which have long extensions.

14.3   PAIR-GAIN SYSTEMS

Until 1981, the only systems that fell into this category were the analogue subscriber carrier and an old and little used "line connector 1A" – a rudimentary Strowger concentrator. Inexpensive microprocessors and MSI circuits have since resulted in the development of sophisticated analogue and digital concentrators and multiplexers. In some countries these are used extensively to augment existing cables, or to reduce the size of cables that feed remote communities of homes. There has been a lower utilisation of pair-gain systems in this country, because lines are comparatively short and therefore relatively inexpensive.

14.3.1   Analogue Subscriber Carrier (1+1)

These systems provide a 'carrier' derived telephony circuit in addition to the 'physical' baseband telephone circuit on a single exchange line, the two circuits being separated by filters. At the exchange, the two exchange circuits feed into the carrier equipment which is connected to a single local-line pair. At the subscriber end of the line, a low pass filter is situated at the teeing point (usually a Distribution Point) where the two circuits are separated – a normal baseband but filtered circuit to one customer, and a carrier derived circuit to the other. The carrier equipment is situated in the customer's premises.

Several techniques can be used for the carrier modulators and demodulators – direct AM, AM with IF's and narrowband FM. The subscriber carrier systems employed in the UK network utilise the first two techniques with double sideband amplitude modulated carriers at 40 kHz and 64 kHz. The carriers are multiples of 4 kHz in order to prevent beating with spurious tones from FDM and PCM junction equipment.

A block diagram of the UK system is shown in Fig 14.3 and further details of the system are given by Kingswell and Toussaint (1). The system will tolerate losses up to 40 dB at 64 kHz – that is, a loop

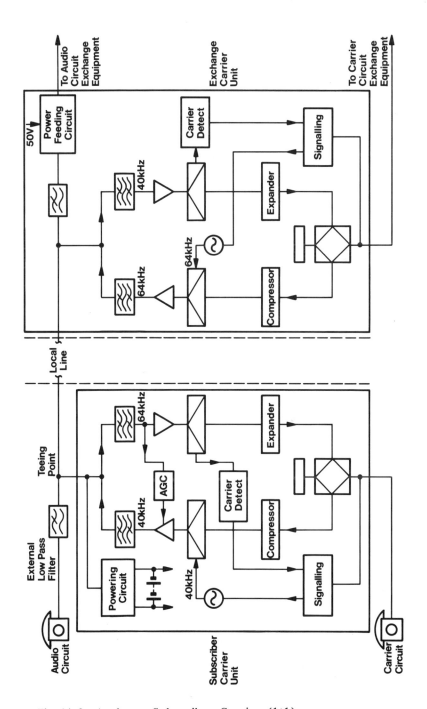

Fig 14.3   Analogue Subscriber Carrier (1+1)

resistance of approximately 1000 ohm. The exchange unit of the system transmits the 64 kHz modulated carrier to line at a level of 0 dBm; the subscriber unit has a variable transmit level that is controlled by the level of the incoming 64 kHz carrier, so that the 40 kHz carrier received back at the exchange unit is always at approximately the same level of -35 dBm. Where the received signal levels at the exchange vary considerably, any crosstalk between one pair and another is worsened by the difference between the levels. Restriction of the range of the received signal level, termed cross-coupled AGC, stops this problem as at any point on the line, the signal level of the customer-to-exchange carrier is approximately the same.

Signalling of loop disconnect pulses is transmitted by the simple presence or absence of carrier, and ringing by the modulation of the carrier by the ringing voltage. When the system is not in use, the carriers are switched off and the system powers down. The signalling system has only the capacity for two states in each direction, and restricts the use of the carrier line to basic telephones; it does not cater for coinboxes, subscriber private metering, earth loop calling etc which require other signals to be passed over the circuit.

When first introduced into the network, subscriber carrier suffered greatly from radio interference, as its main application is on long length lines that are difficult to relieve, and which often have long sections of overhead cable. The development of single chip compandors (compressor/expander) which restrict the dynamic range of the carrier has almost eliminated the problem.

The subscriber unit is powered from a Nickel-Cadmium battery that is trickle-charged over the line when it is not being used by the baseband subscriber (Freer and Matthews (2)). A 15 V, 225 mAH battery supplies power to the subscriber unit; trickle charging at 12-16 mA gives the carrier system the capacity of handling up to ten times the average calling rate on both the carrier and baseband circuit.

Finally, the design of the low pass filter at the teeing point deserves a special mention. When first introduced, the filters were designed around 600 ohm impedances, the nominal impedance of the exchange and the local line. However, telephones are designed to match into real lines, and the use of such filters causes high levels of sidetone for the baseband subscriber. The teeing point filter was therefore redesigned to give a better match to the telephone by the addition of resistance in the filter; a greater insertion loss results, and a suitable compromise has to be sought between sidetone and attenuation.

14.3.2   Multiplexers

The main application of multiplexers at present is as customer equipment connected to digital private circuits which can provide voice and data circuits multiplexed into 64 kbit/s or 2048 kbit/s circuits. Multiplexers are used in the public network but not in great quantities in the UK; their incidence in other countries, especially North America, is much higher.

Analogue multiplexers, or multi-channel carriers, are similar to small FDM systems. The basic principle of operation is very similar to the 1+1 system. Typically 8 channels can be carried on a single pair using double sideband amplitude modulated carriers at 8 kHz spacing.

Digital multiplexers for telephony use are very similar to the PCM multiplexers operating in the junction network between local exchanges.

The signalling cards need to be able to deal with loop disconnect dialling appropriate for the local network, and the detection and application of ringing; a ringing supply must also be provided within the subscriber end of the system which is locally powered.

Although the link between the two ends of the multiplexer are at the primary rate of 2048 kbit/s, some countries have multiplexers at lower 'intermediate' bit-rates, the most common of which is a 10 channel system running at 704 kbit/s. This system could operate over normal twisted pairs. Links at 2048 kbit/s normally have to to be provided over transverse screen cables or optical fibres (see 14.5.1).

Multiplexers will be used extensively in future ISDNs to multiplex a number of 64 kbit/s channels onto a 2048 kbit/s interface. However, they will normally take the form of a multi-line access arrangement for ISDN compatible PBX's (which is therefore a pair-gain arrangement) or as a multiplexer attached to the main exchange and co-located with it. Neither type of multiplexer could be classified as forming part of the local network, although the exchange multiplexer could be placed in the local network and linked remotely. Chapter 12 has more details on ISDN configurations.

14.3.3   Concentrators

Concentrators take advantage of the fact that the calling rate of most customers, in particular residential customers, is very low. Traffic from a group of customers can therefore be carried on a small number of lines by allocating an exchange line to a customer only when required. A switching (concentration) unit, located within the network at a convenient point to serve an appropriate number of customers, is connected over the local network to a complementary expansion unit. This unit, placed in the exchange, feeds the individual customer exchange connections (Fig 14.4). The switching stages can be analogue or digital, and the lines or trunks connecting the two ends of the system may be ordinary local-line pairs, or may be provided by a PCM multiplex system if an even greater reduction in physical pairs is required. Each switching stage is controlled by a microprocessor, and control communication between the two ends is usually by a data link over a separate pair, or a dedicated time slot if PCM is involved. Although calling rates are low, most telephone networks have been designed on the assumption that there is no blocking in the local network; large concentration ratios cannot therefore be contemplated. In addition care must be taken, especially with small concentrators, to make provision for calls between customers on the same concentrator, which tie up two trunks on one call. Some concentrators detect these 'tromboned' calls, and after call set up by the exchange, connect the two customers locally, thereby freeing the trunks to the exchange. The customers' line circuits at the exchange are still busied in order to retain call supervision (metering etc).

Two types of concentrator are presently in use in the UK local network - a large one located at or near the Cross Connect Cabinet and a small one located at the Distribution Point.

14.3.4   Large Concentrator

The large concentrator can serve a maximum of 95 subscribers on 16 cable pairs with a further pair being required as a data link between the two ends of the system. The switch matrix consists of two-stages

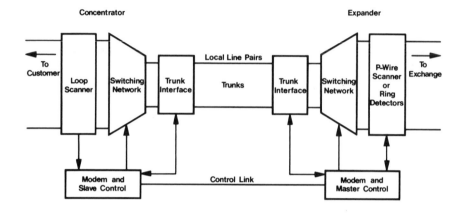

Fig 14.4   Block diagram of a Concentrator System

and utilises a latching crossbar matrix, which once latched, requires no power to keep the connection set up.   The remote end is mounted in a street cabinet close to the existing cabinet, and powered by a battery, trickle charged over the control pair and any trunks not in use.   The concentrator is available in two forms – one which has a standard two-wire connection to the exchange, the other which also accesses the P-wire (ie a 3 wire exchange interface).   This latter version does not require ringing detection, or monitoring of the loop at the customer's telephone, apart from initial call set-up, and is the one used in the UK.

Customer loops are detected by a loop scanner, and a request then made by the slave processor over the 600 baud control link for a free trunk.   The master processor allocates a free trunk after having tested it, via the trunk interface, for short circuit, open circuit, and leakage, before setting up the switch matrices as appropriate.

Incoming calls follow a similar procedure where the P-wire changes condition from 'free' to 'busy'.   The end of the call is also detected by a reversion in the state of the P-wire, but unless the trunk is required for another calling customer the link to the exchange will remain set up.

A disadvantage of using the P-wire to signal customer activity is that there is no differentiation between the 'Parked' condition and the end of call.   Problems can arise from this, when, owing to a cable fault, a large number (17 or more) permanent loops are generated by customers.   When the 'Park' condition subsequently causes the P-wire to become 'free', the concentrator concludes that the call has terminated, and the next looped customer line is connected to the exchange.   Up to 16 exchange registers are thereby permanently busied. The concentrator must therefore recognise the 'Park' condition, and this is achieved by ignoring the P-wire after a loop and then an associated free condition have been present on a P-wire for a certain length of time.

14.3.5   Small Concentrator

BT has developed a small line concentrator (Knox (3)). It concentrates 14 customers onto 4 trunks, and is designed to relieve pair shortages in the distribution part of the network ie between the Distribution Point (DP) and the Cross Connect Cabinet.

The basic design follows the generalised arrangement shown in Figure 14.4. The switching matrix is composed of high voltage semiconductor crosspoints capable of withstanding the harsh electrical environment of the local line network. The remote end of the system is housed in a small waterproof enclosure that can be mounted at the top of a pole. This concentrator is controlled by an associated expansion device at the exchange. CMOS microprocessors at both ends of the system control the set-up of the crosspoint matrices. The processors communicate over a 300 baud full duplex control channel carried by a fifth trunk, which also supplies a constant 30 mA current feed to power the remote concentrator.

The switching matrix is a 16-line to 4-trunk arrangement. Two ports are therefore spare; at the remote end of the system one port is allocated to a loop detector which, by setting the appropriate crosspoints, scans all the customers' lines. The second port is allocated to testing.

The expansion device at the exchange is rack mounted. A 3-wire interface to the exchange is used, ringing being detected from the P-wire. The exchange end also carries out periodic checks for correct operation of the system and reports faults. A VDU can be connected to the exchange end of the system via an RS232 interface to check status and to analyse the performance of the equipment.

14.3.6   Digital Concentrators

Purely digital concentrators are also available. Per-line codecs and digital switching concentrate traffic onto one or more primary PCM line systems. These systems can be very sophisticated and serve a large number of customers. One system, for example, serves up to 512 customers on four, 2.048 Mbit/s links (Campbell and Klodt (4)). These customers can be distributed on up to 8 remote concentrators connected to a single exchange unit, and loss of some of the digital links causes traffic to be reassigned to the remaining free channels. The systems are designed for countries with fairly substantial communities isolated from the main population, and therefore have not so far found widespread application in the UK. The economics of digital concentrators are substantially improved, however, when concentrated traffic can be interfaced directly into the local exchange. In this case there is no expansion stage and the remote concentrator is therefore an outstationed part of the exchange.

A disadvantage of all multiplexers and digital concentrators is that the normal local-line signals have to be converted into a different form - bit patterns or carrier levels. Signalling conditions for various customer apparatus eg coinboxes, subscriber private meters, are not internationally standardised, nor is it usually economic to equip each system with the facilities to transmit signals appropriate for the whole range of customer apparatus. The result is that the systems are inevitably restricted to operating with the more common types of apparatus.

Analogue concentrators do not need any signalling translation once a path has been set up, and therefore have the advantage of greater compatibility with customer apparatus.

14.4 NEW SERVICES

The desire to increase the use of the telephone network so as to recoup more of the administrations' investment has led over the years to a proliferation of non-telephony services eg datel, Prestel, facsimile etc. These Value Added Network Services (VANS) are carried over the PSTN and need no additional network equipment. The local network is the most lightly used part of the network, and some services are being introduced which relate to it particularly. These are not switched through the PSTN, and allow normal telephony service to be provided uninterrupted on the local-line pair. Additional equipment is required at the exchange.

14.4.1   ABC Alarm Service

The first systems entered public service in East Anglia in 1979. Customers are offered equipment to transmit manually triggered fire and police alarms, or automatic alarms from privately supplied equipment, from their premises to a central control point ie police operations room, or fire brigade headquarters. The system is shown in schematic form in Fig 14.5 and described fully by Kingswell and Chamberlain (5). The alarms are transmitted on a high frequency carrier over the local line so that there is no interference with normal telephone services; a local processor at the exchange collates and checks inputs from up to 448 protected premises. From the local processor, information is passed over dedicated circuits to a central processor situated at a suitable exchange, and then to the police or fire control points. The central processor is responsible for routing the incoming data from the local processor and forwarding it to the appropriate display and print unit at the control point. Line and equipment fault reports are also passed to a display and print unit at a BT maintenance centre. Successful reception of the alarm is signalled back to the local processor. Up to 30 local processors can be controlled by a central processor, and up to 15 display and print terminals accommodated. The system can also cater for the connection of a number of local miscellaneous alarms eg exchange alarms.

The customer's equipment consists of a continuously modulated carrier, modulation being controlled by the state of the alarm contacts. The system is powered by the same method as that used in the analogue subscriber carrier, and the alarm equipment and telephone circuit similarly isolated from each other by a low pass filter. On recognising the alarm, the local processor sends back a signal to the customer, where, in the case of the fire alarm button, an indicator is lit; this facility can also be provided if necessary for the police call button. The carrier is continuously monitored at the exchange for the alarm conditions, and also the presence of the carrier; a line fault signal is generated on carrier fail and passed on via the central processor to the BT maintenance centre and the Police Control Room.

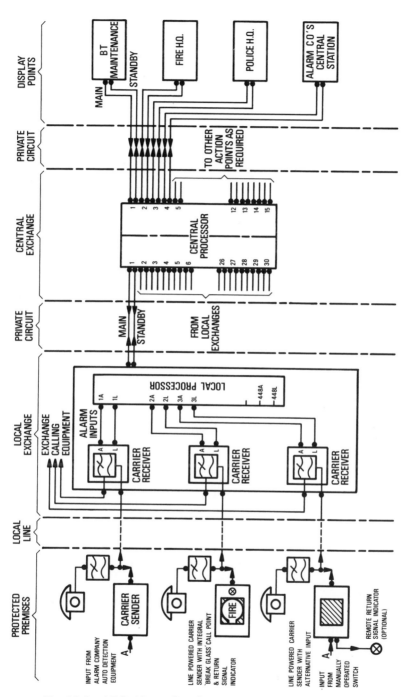

Fig 14.5 ABC Alarm Service

For manually triggered alarms, the carrier equipment is contained in a wall-mounted plastic case with either an integral fire alarm button, or a separate police call button which is installed remote from the carrier sender. Where the system is used with privately provided alarm signal generators, the carrier sender is mounted in a steel case, which has protection against malicious interference by various alarm devices.

14.4.2   Idle Line Utilisation

The average residential customer makes and receives four, four minute calls per day. The line is therefore idle for almost 98% of the time. Assuming that connection to the telephone network can be quickly restored when required, this idle period may be utilised to provide the customer with other services. The use of the idle line in this way was perceived by BT and patents taken out in 1978. Systems based on idle line utilisation are particularly well suited to services which are not time-critical, such as remote utility meter reading and load management (gas, electricity, water). They can equally well be used for any service which involves data transfers between central computers and large numbers of outstations eg alarm/security applications, point of sale services, stock enquiry, viewdata and database access for home computers. (Matthews and Sturch (6)).

BT first began trials on a system called CALMS (Credit and Load Management System) in conjunction with the South Eastern Electricity Board in 1981. This system was developed for use with electricity meters, but it served as a demonstration of the capabilities of idle line utilisation which could be further exploited for other services. Subsequently, BT has funded the development of Bitstream, based upon CALMS, which is due to enter pilot service later in 1986.

The Bitstream system comprises three main elements:

Customer premises   -   Outstation Communications Controller (OCC)
Exchange site   -   Line Concentrator
                          Data Concentrator

A block diagram of the system is shown in Fig 14.6.

OCC

The OCC is inserted into the normal telephone line, and in its quiescent state draws very little current (< 15 µA). It interfaces via V24 ports (variants of the OCC can handle up to 16 ports) to the utility or customer provided terminal equipment (outstations). The OCC, which powers up on receipt of signals from the outstation or the exchange, provides for data communication between the outstation and the line concentrator at the exchange at speeds of up to 1200 baud half duplex.

Line Concentrator

Incoming telephone lines are diverted through the line concentrator which contains a relay allowing the line to be connected to the Bitstream system or the telephone exchange. It detects ringing and loop conditions on the line, controls the switching of the access relay and sets up connections to the data concentrator. A line concentrator can interface to up to 1000 lines.

Data Concentrator

A Local Area Network (LAN) provides message switching between a number of elements connected to the Data Concentrator. Principally, it connects a number of line concentrators (up to 24) to a Data Controller, and the Data Controller to Central Computers (operated by say the utilities), via the Packet SwitchStream network. Line concentrators can be connected to the LAN via modems enabling them to

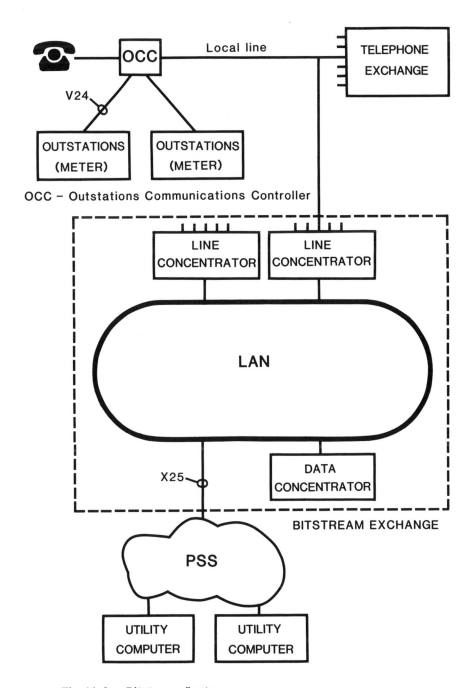

Fig 14.6   Bitstream System

be remotely located to provide economic service to small local exchanges.

The Data Controller provides message handling between the central computer and the OCC, downloads billing information to a central billing system and provides a man-machine interface.

Layered protocols based on the ISO OSI model are used throughout the Bitstream system which can support the following service features.

- single and multiple messages to and from individual outstations
- broadcast of single messages, and collection of responses
- prioritisation of messages
- automatic addressing of messages from particular outstations
- re-routing of calls to central computers.

Bitstream equipment has been designed to be compatible with all existing analogue exchange types in the UK and can support the service on analogue lines served by digital exchanges. It is expected that digital OCC's will support the connection of Bitstream outstations where ISDN is used.

14.5   DIGITAL TRANSMISSION IN THE LOCAL NETWORK

Digital data has been transmitted over the local network for many years, but until relatively recently the data rate was restricted to that which could be transmitted in 300-3400 Hz channels of the PSTN using modems at each end of the circuit. No additional electronics were required in the other parts of the network.

Higher data speeds (made possible by the digitalisation of the main network) and different switching techniques for data (eg PSS) have necessitated the installation of electronics either at the local exchange or within the network itself, as well as in the customer's premises. The techniques used for data transmission in the local network are dependent on the speed required which is again dependent on the service. Speeds and services can be divided into three basic categories:

1   High capacity systems (384 kbit/s upwards). As multi-64 kbit/s channel systems, they can be used to provide a mixture of voice and data services to PBXs, remote multiplexers or concentrators. Some services require higher, non channelised data rates eg video-conferencing and digital video services, high quality sound, LAN interconnections etc.

2   Basic access systems (64-144 kbit/s) for ISDN, digital 1+1 and private circuit data.

3   Low speed data-over voice (2.4 - 19.2 kbit/s) for access to PSS and ISDN. Although these systems have found application in conjunction with PBXs including over external extensions, they are not yet in general use in the PSTN.

Reference 8 (Dufour) describes all three types of system.

14.5.1   High Capacity Systems

High capacity systems cover a wide range of speeds from 384 kbit/s to 8 Mbit/s or higher. A variety of transmission media are used - twisted pair, coaxial cable or optical fibre.

There are obvious economic benefits in using existing twisted pair cables as much as possible. However, these cables are not designed for high frequency use, and they are limited in performance by crosstalk and attenuation characteristics, as well as by impulsive noise from adjacent pairs carrying normal ringing and loop disconnect dialling signals. It is possible to use existing pair type cables for intermediate bit-rate systems running at speeds of 384-768 kbit/s, and even 2.048 Mbit/s systems can be installed on very short lines of up to 1 km in length. However, 2.048 Mbit/s circuits are usually provided over transverse-screen cables, and the use of the existing network is to be regarded as an expedient measure.

Transverse screen cables have an internal screen which separates the pairs into two groups for 'go' and 'return' signals, thereby eliminating the major constraint of near-end crosstalk. Cable fill is 100%, range is up to 2 km between regenerators (0.6 mm copper conductors are used) and there is no impulsive noise from crosstalked telephony signalling.

Transverse screen cables are presently available in three sizes, 80 pair (40 'go' pairs + 40 'return' pairs), 40 pair and 20 pair. Three cable arrangements used are illustrated in Fig 14.7.

(i)   Direct cabling. A small 20 pair cable directly connects each customer's premises to the exchange. This is best used for situations where little or no growth is anticipated.

(ii)   Indirect cabling via a flexibility point. A large cable say, 80 pair, connects the exchange to a flexibility point, with 20 pair cables radiating out to serve individual customer premises. Regenerators, housed close to the flexibility point, can be connected into circuits as required.

(iii)   Tapering cable scheme. This is mainly used where growth is expected. Progressively smaller cables are used on the route out from the exchange with 20 pair cables linking the customer to the main route.

Transverse screen cables are a cost-effective way of providing 2 Mbit/s access. However, they are not easily adapted for operating at higher speeds, and the cost of optical fibre systems is falling, so that future systems will increasingly use optical fibres as the transmission media.

Optical systems are already used in the local network, albeit their penetration is relatively low. Currently, they mainly use the same components that form the main trunk and junction network transmission system. However, cost-reduced systems designed specifically for the short distances encountered in the local network are becoming available.

Fibre systems are generally used for
(a)   high speed connection to the higher order digital multiplexer rates (eg 8 and 140 Mbit/s) that form the main trunk network, for a high speed long distance private circuit;
or (b)   short distance customer-to-customer private circuits, either at CCITT rates or sometimes at other rates for particular applications eg a 10 Mbit/s Ethernet link.

Fibres are also used for cable TV distribution in the local network. The equipment developed by British Telecom for switched star cable TV systems, which multiplexes four 8 MHz TV circuits onto a single fibre, can also be used for private circuit video, or for the transmission of 2 Mbit/s or 8 Mbit/s data in each video channel.

Coaxial cables, in addition to their use for cable TV systems and video, can also be used for the transmission of data. Coaxial cables are used in Local Area Networks, and broadband LAN systems employing cable TV technology have the capability of providing data

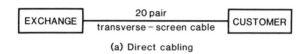

(a) Direct cabling

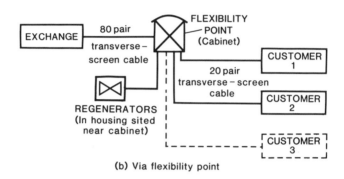

(b) Via flexibility point

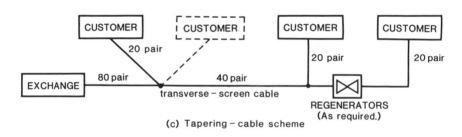

(c) Tapering – cable scheme

Fig 14.7   Transverse-screen Cable Configurations
(From British Telecommunications Engineering)

transmission over a wider area to form Metropolitan Area Networks. Packet switching techniques are employed to switch generally low speed data (up to 19.2 kbit/s from terminals connected via V24 interfaces) over high speed backbone links on the coaxial cable running at 128 kbit/s to 5 Mbit/s.   Such networks are currently under evaluation and similar techniques but using fibre technology may form the basis of multi-service networks in the future.

14.5.2   Basic Access Systems

A variety of techniques can be used for basic access systems in the range 64-144 kbit/s.   Four-wire baseband transmission systems have been used for a number of years for private circuit data at 48 kbit/s and 64 kbit/s.   In order to minimise costs, it is essential that transmission is contained to a single pair of wires.   ISDN has been the driving force behind the development of two-wire basic access systems.

Burst-mode and echo-cancellation are the methods that have been subject to the most investigation, and echo-cancellation has emerged as the preferred technique for two-wire transmission in the local network. These techniques are fully described in Chapters 4 and 12.

Briefly, each end of a burst-mode system alternately sends bursts of data at a sufficiently high speed to allow for the time dividing of the two directions of transmission and for the delay in transmission down the line (Fig 14.8). An early system used for the pilot ISDN service has bursts of data each 20 bits long (excluding synchronisation) with a repetition rate of 4 kHz and a bit-rate within the burst of 256 kbit/s. Burst-mode systems are relatively easy to implement, and avoid near-end crosstalk. However, because bit-rates must necessarily be higher than baseband systems, the attenuation limited range is lower than echo-cancellation systems.

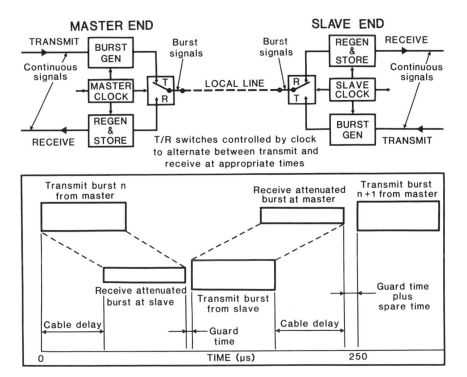

Fig 14.8   Burst-mode System

Echo-cancelling hybrids (Fig 14.9) model the echo response of the local line. A normal 2W-4W hybrid provides some separation of the two directions of transmission. The echo-canceller (which is adaptive) takes as an input the last few digits of the data sent to line and produces an output which is estimated to be the line echo at the instant of receive signal sampling. The true receive signal is produced after subtraction of the echo estimate and output of the hybrid.

An 88 kbit/s echo-cancelling hybrid utilising an analogue transversal filter is used in the pilot ISDN service. Other echo-cancelling hybrid systems are under development for 144 kbit/s access, and echo-cancellers are likely to be the predominant type of basic basic access transmission system in the future.

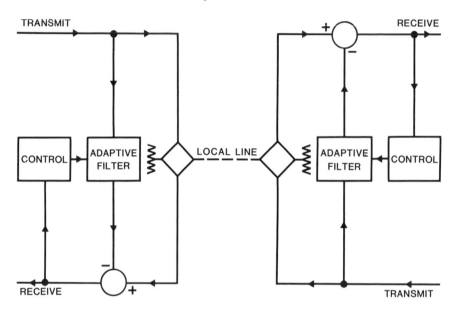

Fig 14.9   Echo-cancelling Hybrid System

14.5.3   Digital 1+1

Echo-cancellation techniques are also employed in a digital 1+1 system that is used both as a replacement for the analogue 1+1 system and as a local line transmission system for 64 kbit/s private data service.   The system runs at 96 kbit/s with 8 kbit/s used for synchronisation and housekeeping functions and 8 kbit/s for signalling. One channel at 64 kbit/s is available for customer use and there are two spare 8 kbit/s channels.   The WAL2 line code is used.

Filters separate the digital transmission signals from the baseband analogue signals which include the normal telephony signalling conditions – ringing and loop disconnect dialling.   Filter design requirements in terms of distortion and inter-symbol interference can be eased by the use of line codes with little low frequency energy.   In this respect therefore, the WAL2 line code used in the digital 1+1 has an advantage over codes such as AMI and 4B3T even though the latter have better performance in terms of reach and crosstalk.

The filters themselves perform three functions:

(i)   Stop impulses and signalling activity on the baseband circuit from causing excessive errors on the digital circuit.

(ii)   Limit the noise on the audio circuit produced by the bottom part of the digital line code spectrum.

(iii)   Attenuate above audio signals on the baseband circuit, again from the overlap of the digital spectrum so that no interference is caused to FDM and PCM systems when the line is connected through the exchange to junction circuits.

14.5.4   Low Speed Data-over-voice Systems

These systems are used mainly on PBX extensions including external extensions as a cost-effective way of providing simultaneous data (up to 19.2 kbit/s) and voice over a single pair of wires. Analogue systems are most often used and are generally variations on analogue subscriber carrier systems with either straight amplitude modulation of the carrier by the data, or frequency shift keying.

Digital systems using burst-mode techniques can be used for low cost access to ISDN.

14.6 CONCLUSION

The last few years has seen a considerable increase in the colonisation of the local network by electronic systems. It has been a challenge to design systems to operate successfully in the rigorous conditions imposed by the local network in terms of interference, physical environment and compatibility with existing equipment. This challenge has been met and overcome. Techniques and equipment have been developed that successfully exploit the existing pair-type local network to provide services and an information capacity far beyond that originally envisaged. Although a little more can always be wrung out of the copper network, attention is being increasingly focussed on other transmission media, in particular optical fibres, that can provide an almost unlimited bandwidth. It is these systems which are likely to occupy the designers and planners of electronic systems for the local network in the future.

REFERENCES

1   Kingswell, L.W., and Toussaint, G.C., 1972, 'A Subscribers' Carrier System for the Local Network', POEE Journal, Vol 64, Part 4.

2   Freer, B.R., and Matthews, O.C., 1973, 'Local Battery Charging over Subscribers' Lines, POEE Journal, Vol 66, Part 2.

3   Knox, D.M., 1984 'Semiconductor Space-Switch Concentrators in the Local Network', British Telecom Technology Journal, Vol 2, No 1.

4   Campbell, D., and Klodt, R., 1980, 'Subscriber Carrier Evolution - DMS-1A (2 Mb/s)', Proc. ISSLS 80.

5   Kingswell, L.W., and Chamberlain, I.C,, 1979, 'Alarms by Carrier', POEE Journal, Vol 72, Part 2.

6   Matthews, O.C., and Sturch, P.M.E., 1977, 'Remote Electricity Supply Meter Reading and Control', Proc. 3rd International Confidence on Metering Apparatus and Tariffs for Electricity Supply.

7   Vogel, E.C., and Taylor, C.G., 1982, 'British Telecom's Experience of Digital Transmission in the Local Network', <u>Proc ISSLS 82</u>.

8   Dufour, I.G., 1985, 'Local Lines – The Way Ahead', <u>British Telecommunications Engineering</u>, <u>Vol 4</u>, <u>Part 1</u>.

Chapter 15

Radio

Bob Swain

15.1 WHY RADIO?

Traditionally the transmission needs of the local
network have been met by copper pair and coaxial cable for
a service overwhelmingly dominated by telephony.
Furthermore when installing cable it has been wise to plan
for growth on the basis that every factory, office and
nowadays home will require a telephone service.
Consequently the present custodians have inherited a
robust, extensive but relatively low-cost local network
right up to the customer's premises. However, the moderate
demands made by telephony on transmission performance have
resulted in a network which will require some selection of
pairs, adjustments, modification and equipping with
apppropriate terminal apparatus if it is to accommodate
some of the more modern communication services. For
example, voice-band data ranging from a few bit/s through
to 64 kbit/s pcm to the integrated services digital network
(ISDN) rate of 144 kbit/s, and slow-scan television. To
provide digital services with transmission rates in excess
of 2 Mbit/s (or equivalent analogue signals) is also beyond
the scope of much of the existing network. So special
arrangements must be made for them. What are these special
arrangements? The conventional engineering view would say,
unequivocally, co-axial cable or optical fibre laid direct
to the customer's premises. But in an age of rapid
evolution, if not revolution, of new telecommunication
services and facilities, and a national policy of
liberalisation and competition, the marketing manager's
view may call for a somewhat different solution. His
objectives would likely be; rapid and timely provision of
service at a cost acceptable to the market and/or customer
as well as his own balance sheet, and with a potential to
recover a large proportion of the investment on cessation
of service (especially important for temporary services).
These objectives can be easily met if the customer is
already served by, or close to, a modern high-capacity
transmission cable, but this is not yet the general case.
The cable solution can easily incur the costs of, at worst,
laying cable in duct through a major conurbation to
providing, at best, an overhead drop-wire service.

It is with this background that much effort is being expended in providing a radio transmission option. Many manufacturers currently provide microwave radio equipment for this market. Such systems offer many advantages, for example;

.   quick installation,

.   reasonable cost for low-density and/or special services,

.   all equipment can be readily recovered for use elsewhere,

.   good area coverage up to visual range (~10 km) from a suitably high vantage point,

.   point-to-point service without inconveniencing other city dwellers,

.   radio equipment can be designed to cater for a range of signals without excessive cost penalty.

The disadvantages include;

.   radio is prone to interference from inside, or outside, the system,

.   radio signals can fade due to the weather,

.   radio generally requires line-of-sight between terminals although mobile radio communications is an exception,

.   radio spectrum is a limited resource and so the transmission capacity is limited in any given area.

The marketing man, if he is really abreast of the times, will have also detected a burgeoning need from large sections of the community for communications while on the move, either in a car or on foot. This market ranges from the needs of company directors and service/sales engineers, where avoidable travelling is wasted money, to the consumer electronics end of the market typified by the cordless telephone bringing with it the freedom of movement in home, factory, office and store. He, marketing man, will have realised that only radio can service this untapped but potentially enormous market.

In the rest of this chapter examples are given of both uses of radio in the local network;

.   the relatively new point-to-customer service, and

.   the mobile service.

Firstly, however, it is pertinent to quickly review the radio carrier frequency and characteristics of fading appropriate to each service. The many other, equally important, matters of detail system design and construction must be considered beyond the scope of this book; it will only concentrate on the important uses, characteristics and problems of each service and tend to look forward rather than backward. Nevertheless, for background, the important parts of any radio communication system are indicated in fig 15.1.

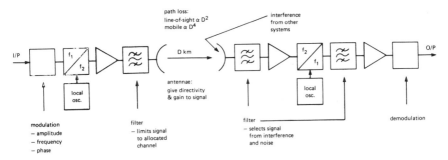

FIG 15.1   Basic Radio Channel

15.2   FREQUENCY & FADING

15.2.1   Frequency Allocation

The International Telecommunications Union (ITU), through the work of the World Administrative Radio Conferences, has designated the uses of the radio spectrum from 9 kHz to 275 GHz. The uses cover the needs of amateur radio enthusiasts, radio navigation, satellite telecommunications, maritime, aeronautical services, terrestrial and space research, inter-satellite links, as well as the two categories of interest to this chapter, viz fixed point-to-point telecommunication links and land mobile communications. Although the ITU cannot enforce its internationally agreed recommendations most countries follow them to ensure that spectrum management can be practised with a view to minimising mutual interference between systems sharing the same frequency band. This process is known as co-ordination and in Europe means co-operation between the various countries on the exploitation of the scarce resource. Co-ordination is also vital within a country if the spectrum is to be effectively used to the benefit of all.

Point-to-customer services are operated in the
microwave bands above 1 GHz and systems are known to
operate at frequencies around 10, 13, 15, 19, 21 and
29 GHz, with further developments in train at 39, 51 and
60 GHz. On the other hand recent land mobile radio systems
operate below 1 GHz sharing the band with broadcasting;
police, fire and ambulance; maritime; aeronautical; and
radio navigation services. Much of the UK spectrum
allocation to radio telephone systems lies between 100 and
174 MHz with cellular mobile radio systems operating betwen
890 and 960 MHz. However, the UK situation is changing
because the Merriman Committee in its Independent Review of
the Radio Spectrum (30-960 MHz) (1) recommended that the
405-line television signal bands 41-68 MHz and 174-225 MHz
be withdrawn from broadcasting use and re-allocated
primarily to land mobile services. Even so, such a major
infusion of spectrum barely matches the expected growth to
the year 2000.

15.2.2 Radio Fading

The two types of local network radio differ in their
fading characteristics and what is more differences appear
between rural and city based systems. Thus it is not
possible to produce a unified expression, or model, that
relates propagation fading to the necessary, or achievable,
operational range or outage criteria. Each case must be
treated separately.

15.2.2.1 Point-to-Customer Microwave Links

Microwave signals can be impeded by the effects of
rain or bulk changes in the atmospheric refractive index.
As indicated in fig 2 refractive index variations give rise
to multiple-path transmission and the net result is signal
cancellation which, in wide bandwidth systems, is
accompanied by distortion of the amplitude and
phase-frequency transmission characteristics. The
phenomenon generally applies only to links longer than,
say, 5 km and only in rural areas. The turbulent air
generated in cities (eg rising hot air from buildings)
destroys the stable meteorological conditions implied in
fig 15.2. Unfortunately an essentially similar effect can
result from signal reflections from buildings.
Consequently multipath fading does occur in cities but in
this case the reflections are permanent, not fleeting, and
so can be observed during setting-up of a link. Corrective
action takes the form of realigning the antenna bore-sight
bearing.

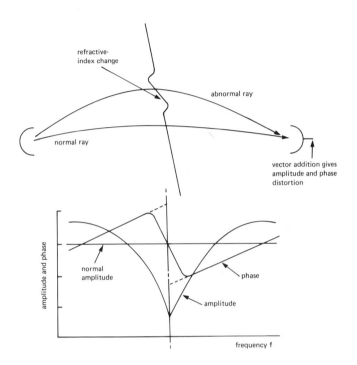

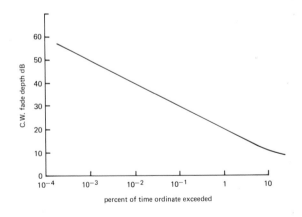

FIG 15.2 Multipath Fading

Refractive-index changes can take its toll at any frequency but in practice excess attenuation due to rain dominates for systems operating above about 12 GHz. At this frequency attenuation is due to energy absorption by the rain drops but as the frequency rises towards 100 GHz energy scattering by the droplets predominates - with the result shown in fig 15.3.

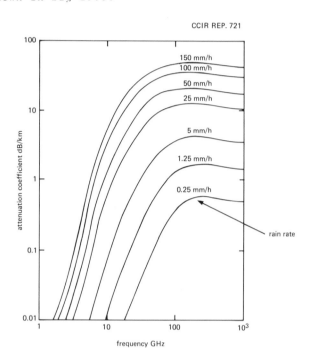

Fig 15.3   Rain Attenuation

The shapes of the droplets also cause vertically polarised radio waves to be partially converted to horizontal, and vice versa, fig 15.4. This phenomenon holds important implications for systems that, in an endeavour to achieve good spectrum utilisation, re-use frequencies on orthogonal polarisations. Clearly rain is also no respector of rural or urban terrain.

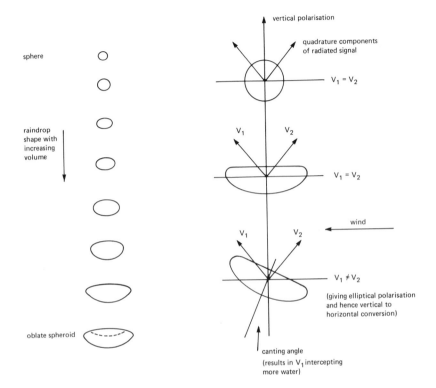

FIG 15.4   Rain Depolarisation

Both propagation phenomena, taken together with equipment reliability, set the limit of availability in a radio system. But, unlike equipment failure, system outage due to fading is self healing and protection against it is afforded by excess tranmission gain; a fade margin implemented by automatic gain control.

There is another source of excess transmission loss, particularly relevant to future systems, namely gaseous absorption. Fig 15.5 shows a number of absorption bands, but that due to oxygen at 60 GHz is currently the most important. In this case the loss is stable, and at 16 dB/km a significant factor of a radio system's transmission equation.

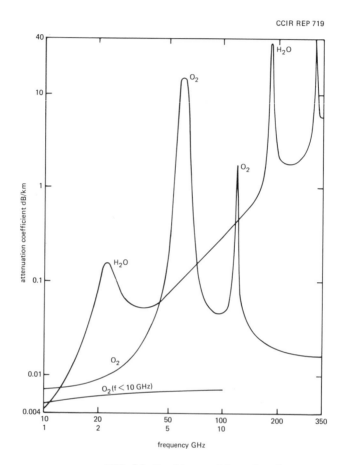

CCIR REP 719

FIG 15.5   Absorption Bands

15.2.2.2   Mobile Communication Systems

These systems at present operate below 1 GHz and have the largest percentage of their customers in towns and cities, with equipment either installed in vehicles or hand-held and thus free to go into buildings. Generally speaking the central base-station antenna is not placed dominantly high with respect to local buildings and hills, and certainly the mobile receiving and transmitting antennas are well buried amongst the buildings and adjacent vehicles. Line-of-sight communication cannot be assumed and consequently propagation characteristics differ markedly from those for point-to-customer microwave links.

Since buildings, trees, hills etc obstruct the radio
path then we must expect signal loss and cancellation due
to;

blockage by obstacles hence leading to absorption and
scattering of signals,

diffraction around obstacles, and

multipath transmission due to many signals reflecting
off buildings before arriving at the receiver thereby
causing rapid signal fluctuation.

These factors cause wide signal variations over the
intended coverage area and can only be described in
statistical terms, but in brief are as follows.

Mean received power (P_r) decays in proportion to the
inverse fourth power of distance (D).

Variation of P_r (averaging distance 20m) about the
mean is a log-normal distribution, with a standard
deviation between 6 and 10 dB.

Instantaneous received signal strength rapidly
fluctuates due to multipath over distances the order
of wavelengths yielding a Rayleigh probability
distribution.

Received signal is subject to Doppler shift due to
motion of the mobile receiver.

It is clear that compared to point-to-customer systems
mobile radio transmission is a far more hostile and
unpredictable environment. The effects are
diagrammatically shown in fig 15.6.

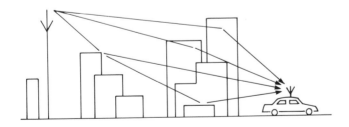

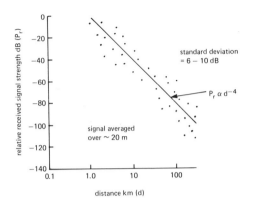

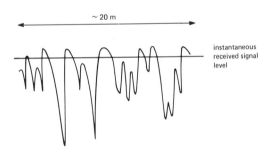

FIG 15.6   Mobile Radio Fading

Although the foregoing has concentrated on external transmission essentially similar results have been seen for transmission within buildings of conventional brick and concrete construction.

15.3   LOCAL NETWORK LINE-OF-SIGHT LINKS

Two types of microwave point-to-customer links are in use or under development, they are;

• conventional point-to-point, and

• point-to-multipoint systems

5.3.1   Point-to-Point Systems

Microwave radio is used for distribution or relaying private or broadcaster's video signals. However the advent of new broadband services (confravision, high-speed data at , 8, 34 and 140 Mbit/s, multiplexed telephony etc) as an increasingly important part of common carrier's publicly offered service is creating a need for exchange-to-customer broadband connections, especially in the urban environment.

Point-to-point links can operate at many frequencies eg. Mohamed and Pilgrim (2)) and are typified by relatively light-weight man-portable equipment that can be easily mounted on a flat roof, or be allowed to radiate through a window for temporary use, and yet has an operational range bordering on 10 km. All this has come about by developments in microwave circuit technology which has led to compact, cost-effective, equipment that can be installed behind a high-gain directional parabolic antenna (40 dB gain, about 0.5m diameter).

Broadband services will be required mainly in commercial and industrial offices which tend to be clustered in towns, cities, and industrial areas. In this comparatively dense radio network, random deployment of links is not a good idea, because, for each proposed new link, interference co-ordination with every other link installed in the locality becomes necessary to avoid mutual interference, and hence the ability to provide service quickly is lost.

Instead, a preferred strategy is a nodal deployment of links around a central node, rather like the spokes of a wheel, as shown in fig 15.7. The system node must be located on a tall building having a flat roof suitable for accommodating several transceivers simultaneously; good access, availability of a secure power supply, and a safe working area with low vandalism risk are highly desirable. The location of the node must ensure unobstructed line-of-sight to the majority of potential customer sites in the locality, and furthermore must be selected with a view to achieving a more-or-less uniform angular distribution of links around it, as far as this is possible by projection of future needs for service.

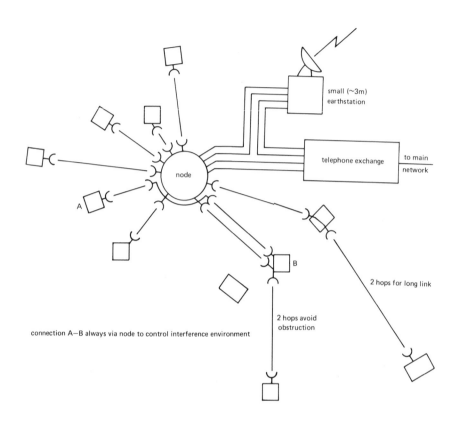

FIG 15.7   Nodal Configuration

Even with a nodal configuration the system must be properly designed if a dense network of radio links is to be assured. Rainfall will cause fading due to intense localised storm cells rather than to more widespread but gentler rain but the predominant factor is differential fading, whereby the nodal receiver on a longish link, suffering rain fading at its outer end, must cope with unfaded interfering signals from shorter links radiating towards the node from all points of the compass. The use of an antenna having excellent front-to-back ratio and directivity aids in overcoming this problem, as does a reduction in transmitter power on the shorter links, which can nevertheless maintain a more than adequate fade margin. By these means a properly designed nodal system allows new links to be installed rapidly, safe in the knowledge that there will be no harmfull interference with existing links in the locality.

These systems are called point-to-point because each terminal is exclusive to one link. There are, however, many present and future examples of common-carrier services in which the transmission signal is digital and ranges between a few kbit/s of data through 64 kbit/s pcm up to the integrated services digital network capacity of 144 kbit/s. Unfortunately low capacities cannot be efficiently conveyed by the point-to-point system for many reasons:

- It is difficult to make spectrally efficient narrow band systems at moderate GHz frequencies; narrow-band radio-frequency filters have high insertion loss and oscillator stabilities become significant with respect to transmission bandwidth.

- It does not make economic sense to burden a low bit-rate data service with cost of a point-to-point link incorporating two transmitting and receiving terminal equipments.

- Low bit-rate services will be on-demand, dial-up, services and thus, cannot afford the time taken to accommodate even the point-to-point system's relatively free and easy radio interference avoiding rules.

To overcome these objections the so-called point-to-multipoint systems were developed.

15.3.2 Point-to-Multipoint Systems

These systems mitigate the above deficiencies by;

- using only one nodal (central station) transmitter to serve all outstations, and

- all outstations are served on a time-division (TDM) basis, and respond in concert with each other by time-division multiple access (TDMA) means.

Thus cost per customer is reduced by the former and transmitted bit rate is effectively increased by the latter. The latter also avoids mutual interference because although all outstations transmit (and receive) on the same frequency they radiate their signals in time sequence, not concurrently. The central station must have an antenna that radiates and receives in all directions so that all outstations are served, the outcome is low antenna gain with a consequent reduction in operational range. Nevertheless point-to-multipoint systems have been, developed for frequencies of 10, 19 and 21 GHz, and at these frequencies operational ranges approaching 10 km can be achieved.

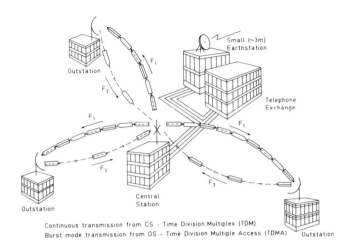

Continuous transmission from CS - Time Division Multiplex (TDM)
Burst mode transmission from OS - Time Division Multiple Access (TDMA)

FIG 15.8   Multipoint System Operation

The operation of the 19 GHz system (Hewitt, Scott and Ballance (3) also Mohamed and Ballance (4)) is illustrated in fig 15.8.   A central station (CS) would normally be located on a tall building with cable connections to a nearby exchange.   At the CS a number of data circuits with rates varying between 12.8 kbit/s and 144 kbit/s are time division multiplexed (TDM) into an 8.192 Mbit/s data stream.   This signal is used for digital modulation of an RF oscillator at approximately 19.6 GHz.   The modulated signal is then transmitted in a continous code from the CS antenna.   Return transmission from identical rooftop radio terminals or outstations (OS) occur at a second frequency around 17.8 GHz.   Subscriber data at up to 144 kbit/s are transmitted at 8.192 Mbit/s to the CS in interleaved bursts with adequate guard time between adjacent bursts.

A line-of-sight path from an OS to a CS is necessary for system access.   However, each OS can serve up to 5 subscribers so that line-of-sight is not essential if there is a suitable building in the neighbourhood as illustrated in fig 15.9.   This method of circuit provision also results in a reduction in the cost of radio terminal per kilobit circuit.

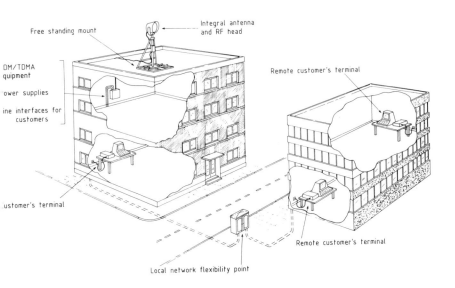

Free standing mount

Integral antenna
and RF head

DM/TDMA
quipment

ower supplies

ine interfaces for
customers

ustomer's terminal

Remote customer's terminal

Remote customer's terminal

Local network flexibility point

FIG 15.9   Typical Outstation Configuration

In the multipoint system the time-division frame
structure is important. Although it is easy to synchronise
outward radiation from the central station, on the return
journey allowance must be made for the different link
lengths if synchronisation is to be achieved at the
central-station receiver. However, it is also possible to
alter the transmission capacity to and from any outstation.
Indeed, in principle, the whole 8 Mbit/s capacity could be
allocated to one outstation, or alternatively an outstation
could switch off.

15.3.3   Absorption Band Systems

Towards the end of the century as demand increases for
broad and narrow-band services, consideration will need to
be given to the intense exploitation of frequencies above
29 GHz. Of particular interest is the 60 GHz absorption
band, fig 15.5, because although the 16 dB/km attenuation
will limit operational range it will also limit the
distance at which unacceptable interference could be
caused. Thus frequencies will be used much more often in a
given area, thus accommodating much more telecommunications
traffic per MHz per km^2.

15.4. MOBILE RADIO COMMUNICATIONS

Mobile communications can either be;

• radio telephone communication to vehicles or people roaming over large areas, or

• cordless telephones to people roaming in the limited environs of office, factory or home.

15.4.1 Radio Telephone Systems

Current commercial mobile telephone systems use conventional analogue frequency-modulated transmission and can be considered as radio extensions to the public switched telephone network (PSTN) with the same range of automatic direct-dialling facilities. Nevertheless there are major changes taking place in this field that bodes well to radically alter the present perception of a radio-telephone as being a rich man's chauffeur-driven telephone to that of a widespread cost-effective means of communication to all who require communication whilst on the move within, or between, cities. it is instructive to note that in 1985 licences to operate new systems included a clause calling for service to be provided to 90% of the population within about 5 years.

Historically radio-telephone engineers have sought to cover the largest possible area with their base-station transmitters by maximising radiated power and choosing the best vantage point for the antenna. Even choosing the highest hill will not guarantee total coverage behind buildings, in tunnels or over hills and these "radio holes" have to be filled by local boost stations, sometimes operating on the same frequency thus causing beat situations. So coverage was patchy and certainly not predictable, but the major problem was shortage of frequencies which constrained the potential market.

This problem was tackled from two angles. First, the obvious solution of allocating more channels. In concert with the rest of Europe the Department of Trade and Industry (Radio Regulatory Division) has made available 1000 extra channels shared between two operating companies. Their frequencies lay in the new 900 MHz mobile radio band; specifically 1000, 25 kHz – wide channnels between 890-915 MHz duplex-paired with 935-960 MHz. This extra allocation was still insufficient and so, secondly, a means of re-using the channels within a given area had to be found. This objective was achieved by confining the radiated signal to relatively small cellular areas about the base station, by adopting low antenna heights and using the propagation characteristics of radio transmission in an urbanised area.

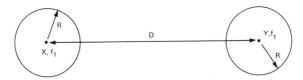

R = maximum operational range
D = re-use distance
f_1 = frequency

FIG 15.10   Channel Re-use

Consider fig 15.10.   The two base-station transmitters (X
and Y) serve their respective cell areas with the same
frequency f_1.   If D is small then the signal from X will
interfere with a receiver in Y's coverage area with a
resulting unacceptable wanted-signal to interfering-signal
ratio (S/I).   It was noted earlier that in the mobile radio
environment received power is proportional to D^{-4}.
Consequently it can be shown that;

$$S/I = \left[\frac{D - R}{R} \right]^4$$

If it is decided to re-use any channel frequency n-times in
a symmetrical arrangement of cells within a major
connurbation, then

$$\frac{S}{I} = \frac{1}{n} \left[\frac{D - R}{R} \right]^4$$

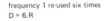

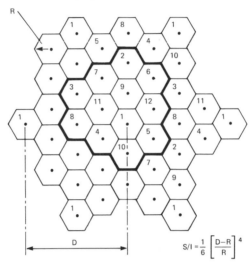

$$S/I = \frac{1}{6} \left[\frac{D-R}{R} \right]^4$$

FIG 15.11   12-Cell Array

It is plain, therefore, that for a given cell radius R and
required mean S/I one can determine D (the frequency re-use
distance) as a function of n. It is possible, therefore,
to draw again the single-dimension frequency re-use model
of fig 1 in two dimensions. The often shown, classical,
example is a hexagonal array of cells, fig 11. For the 12
cell repeat pattern shown D = 4.6R, but in this case D may
be too small and hence intereference too high. With the
12-cell system the number of channels per cell becomes N/12
where N is the total number available. Assuming N/12 is
significant then a trunking advantage can be realised in
each cell and also obtained again when that particular set
of frequencies is re-used. In central London cell radius
is of the order 2 km, whereas in outer London cells are
bigger because traffic density is lower. In principle the
tessellated structure could cover the whole country and
this is the intention of the licence clauses. In practice
cell size varies and most significantly local variations in
propagation characteristics make them highly irregular in
shape.

In system terms, the most notable effect of the
cellular arrangement is that it is now necessary to track
each roaming mobile so that when it leaves one cell it can
be immediately handed over to the next with a consequential
automatic change in operating channel frequency. This
process, known as hand-off, involves considerable
inter-cell control and signalling. It also means that the
traffic statistics of each cell must recognise that calls
are not only generated within its boundary but are also
handed over from, and to, the adjacent cells.

Cellular mobile telephone represents, therefore, a
major evolutionary step in mobile system control and two
such systems started services in the London area early in
1985.

15.4.2 Cordless Telephones

What is a cordless telephone? For our purposes it is
defined as a radio extension-telephone in which the
allocation and use of radio channel frequencies is
autonomously determined unlike a cellular mobile radio
system which is centrally controlled. The market potential
of cordless telephones (CT) was first indicated in the UK
by illicit imports which infringed UK laws, regulations and
practice in many respects, of which the following were the
most significant.

.   The operating frequencies were allocated to maritime
     and broadcast television services.

.   Transmission characteristics did not follow UK
     practice.

.   No dialling security; calls could be made over other
     peoples telephone lines.

This last point was of considerable concern for it
raised the spectre of many thousands of disputed telephone
bills, with consequences to customer confidence.
Consequently the UK radio-regulatory authority produced a
cordless telephone specification with effect from October
1982.

This specification included;

.   Eight analogue-fm duplex-channel pairs operating
     around 1.7 and 47.5 MHz.

.   Line-access security using one of at least 10,000
     handshake codes preset during manufacture.

.   The same code system to be used to selectively call
     the correct handset from the base unit, thereby
     preventing forced eavesdropping.

Transmission performance matched to UK standards.

Instruments to this specification arrived on the market mid 1983, but looking to the future, the specification does not allow the cordless telephone principle to be fully exploited, indeed it has notable limitations. Eight pre-selected channels will in due course restrict the instruments to rural and at best suburban use where CT concentration is lowest. PACTEL International in its report to the Eurodata Foundation on Future Mobile Communication Services in Europe (5) concluded that a 100m-range cordless telephone could generate a maximum user density that, "might be as high as 5000 users/km^2 in major conurbations". There is therefore a potentially large market for cordless telephones that use modern technology to mitigate the intense interference likely to exist in major conurbations where CT density could well exceed 5000 CT/km^2.

15.5 CONCLUSION

The use of radio has been reviewed from the point-of-view of what is possible and how its unique problems can be solved, or at least contained. It is true to say that radio transmission is having something like a revival, not in the traditional long distace trunk and satellite fields but in its use in the local network between exchange (or network node) and the customer's premises or, in the case of mobile radio, the customer himself. Although radio will never equal the inherent capacity of an optical-fibre network (which in time will become all pervasive) it can offer rapid, on-demand, cost-effective, flexible and mobile service; marketable qualities that will see a permanent place for radio in the local network.

ACKNOWLEDGEMENT

Acknowledgement is made to the Director of Research for permission to make use of the information contained in this paper.

REFERENCES

1   Independent Review of the Radio Spectrum (30-960 MHz), Interim Report; Chairman Dr. J. H. H. Merriman, HMSO Cmnd 8666.

2   Mohamed, S. A., and Pilgrim, M.; 29 GHz Point-to-Point Radio Systems for Local Distribution. British Telecom Technol Journal, Vol. 2, No. 1, January 1984.

3   Hewitt, M. T., Scott, R. P., and Ballance, J. W.; A Cost Effective 19 GHz Digital Multipoint Radio System for Local Distribution Applications. Contribution to ICC 84, Amsterdam, May 14-17, 1984.

4   Mohamed, S. A., and Ballance, J. W.; 19 GHz Digital
     Multipoint Radio Systems. ISDN Symposium British
     Telecom Research Labs Martlesham Heath, June 1984.

5   Pactel International; Future Mobile Communication
     Services in Europe. Report to the Eurodata Foundation
     on the systems and opportunities for Services to the
     year 2000, September 1981.

Chapter 16

Overseas practices

Peter Studd

16.1 INTRODUCTION

Practice overseas varies a great deal. In any one place it consists usually of a mixture of the line plant practices introduced by the main advisors to the local Telecommunications Administration and practices introduced to meet local needs.

As the advisers may be French, Swedish, British or American to quote but a few examples, the line plant practices of these countries are often found to have been introduced 'lock stock and barrel' into overseas countries sometimes with little regard to their suitability.

On of the reasons for this is very often the lack of any production facilities locally to manufacture the thousand and one items that go to make up a local telecommunication network. Consequently it is easier to order these up from established suppliers in the advisers country or origin and to ship them in.

In this way we find British line plant practice established in countries in East Africa, French practices established in the French speaking countries of West Africa, North American practices in oil producing Middle Eastern countries and Swedish practices in unlikely places like Nepal.

The economic situation in the country concerned plays a large part in deciding the nature of the line plant practices adopted as also do local weather conditions.

In some countries overseas you find line plant utilising the latest state of the art but more frequently you find techniques that were introduced at some time in the past and which have remained the same ever since. This is partly due to the conservatism of Telecommunications Administrations and partly due to the need to standardise on one type of plant in order to simplify staff training and to minimise the variety of stores items held.

In these ways overseas practices very often mirror practices in more developed countries with the addition of a time warp which takes you back in the time.

16.2   TELEPHONE

16.2.1   Transmission and signalling limits

As in Great Britain telephone line plant overseas is governed by the limits set at any one time to achieve satisfactory signalling and telephone transmitter and receiver operation.

Table 1.16 gives an example of the way these limits evolved in East Africa for direct exchange lines (DELs), i.e. subscribers lines by means of which only one subscribers' station is connected to the exchange.

TABLE 1.16   Transmission and Signalling Limits

		Type of Exchange		
	24 V C.B.	Pre-2000 type non-ballast MAX, and ATE/RAX	2000 type non-ballast MAX, RURAX and GEC/RAX	2000 type ballast MAX, UAX and Magneto
Transmission limit in dBs	6.6	9.4	9.4	10.2
Signalling limit in ohms	400	650	1000	1000

As can be seen from Table 1.16, it includes figures for automatic, central battery and magneto exchange areas.

16.2.2   Magneto Switchboards

The use of magneto switchboards may come as rather a surprise, but these exchanges are still in use in many remote areas where they vary in size from 30 to 100 lines and until 1975 it was the policy of the East African Administration to allow the use of magneto switchboards in large towns and to allow such switchboards to grow until they reached the size of 400 lines before replacing them by automatic exchanges. Multiple type switchboards were used for those exchanges between 100 and 400 lines in size so that the complete exchange multiple could be reached by each individual operator.

This policy was adopted because employment was in short supply and every telephone operator's post represented a job. The manual boards were fitted with dials so that they possessed the capability of trunk dialling, but the subscribers were called by means of magneto signalling, usually by employment of a key which operated a mains driven ringer, but each position was equipped with a magneto ringer for standby purposes.

Local batteries consisting of dry cells were used to provide the transmitter current in the magneto telephone instruments. These were changed periodically by the local technician maintaining plant in the area who was provided with a simple test set for the purpose of testing the local batteries.

The use of magneto switchboards provided a system that could be maintained and operated by local staff without the need for high calibre technical training. This was necessary because it was very difficult to persuade high calibre technical staff to live in remote areas.

16.2.3   Open wire routes

Table 2.16 gives details of the constants and attenuation at audio frequency for open wire pairs at 4 inch spacing in use in East Africa for local plant.

TABLE 2.16   Open wire pairs - 4 inch spacing

| Gauge | Constants per mile | | | | Atten. dB per mile |
	R Ohms	L Henries	G u mhos	C microfarads	
40lb Bz	91	0.00342	1	0.00884	0.46
70lb Bz	52	0.00319	1	0.0095	0.32
40lb CdCu	52.5	0.00342	1	0.00884	0.31
0.080 CuWd	41.6	0.00296	1	0.0097	0.28
70lb CdCu	30	0.00319	1	0.0095	0.22
0.104 CuWd	24.8	0.00281	1	0.0108	0.20

A glance at Table 2.16 shows that the value/mile for both resistance and attenuation reduce in value as we go down the table, and the gauge of wire increases, starting with bronze wire and proceeding through cadmium copper which is an alloy of the two metals that is harder than pure copper to copper weld which is a wire in which steel wire is coated with copper by electrochemical means. This latter is sometimes called copper covered steel.

The materials used for all these wires are chosen to give greater tensile strength than pure copper which is very soft and ductile. This enables longer spans to be erected between poles and also makes the wire less likely to be stolen by local people for use to make ornaments and personal jewellery.

To take the example of a .080 CuWd open wire pair connected to a magneto exchange it is possible to have a DEL up to 24 miles long. In some remote areas such as in the coffee growing areas amongst the foothills of Mount Kilimanjaro lines of this kind of length really do exist.

Table 16.2 does not include galvanised iron wire as this is no longer used for subscribers' lines. However, trunk lines consisting of 200 lb GI or 400 lb GI open wires can still be found in certain parts of Africa.

16.2.3   Party lines

Furthermore there are magneto exchanges in such areas to which party lines are connected. These are single pair open wire routes which run for miles alongside a road or track and to which maybe up to 10 separate single pair spur routes are connected at different points along the length of the backbone pair. Equipped with local battery magneto telephones such party lines represent an economic way of providing telephone service to remote subscribers.

Signalling is by means of coded ringing based on the morse code the handle of the magneto telephone being cranked the requisite number of times.

Such systems, besides providing telecommunications also provide much entertainment, as what could be more interesting than listening to someone elses telephone conversation. The only difficulty is that when you use the telephone you have to remember that everyone is likely to be listening to you too.

It also offers tremendous possibilities for conference calls consisting of all the local farmers in the area.

16.2.5   Subscriber carrier systems

Party lines introduce their own problems, particularly with regard to privacy and with regard to how to automate them in an economic fashion when automatic exchanges replace the magneto switchboards.

The South African Post Office has faced such problems and has introduced a subscriber carrier system called the SOR118 manufactured by STC, Boksburg which provides the answers. This carrier system operates over backbone open wire routes with spurs attached, that were previously party lines, and provides private telephone service which enables each subscriber to be connected to an automatic exchange with exactly the same facilities as others on the exchange.

Such systems can be provided with open wire repeaters if the open wire route is too long. They can also be provided with cable repeaters designed to amplify the signals if they are brought into the exchange by means of an underground lead in cable. Matching units are provided for the interface between the underground cable and the overhead line, divider units are provided at each T-junction and a termination unit at the end of the open wire line.

The frequency plan is so chosen that it fits in with the 3-circuit and 12-circuit open wire carrier systems that the South African Post Office uses on its high overhead open wire trunk routes, without causing crosstalk.

16.2.6   Telephone cables

Table 3.16 gives details of the constants and attenuation at audio frequency for unloaded paper core lead covered telephone cables used in East Africa.

TABLE   3.16   Unloaded lead covered underground cables

DIAMETER IN MM	GAUGE IN LBS	R OHMS	L HENRIES	G U MHOS	C MICROFARADS				ATTENUATION DBS PER MILE		
					TYPE OF CABLE				TYPE OF CABLE		
					PCTD	PCUT	PCQL UNDER 50 PRS	PCQL 50 PRS & OVER	PCTD	PCUT	PCQL
0.4	4 LB	440.8	0.001	1		0.080				3.27	
0.5	6½ LB	270.8	0.001	1	0.075	0.085	0.0825	0.075	2.68	2.75	2.68
0.63	10 LB	176.0	0.001	1	0.075	0.090	0.0825	0.075	2.05	2.20	2.05
0.9	20 LB	88.0	0.001	1	0.070	0.090	0.0825	0.075	1.35	1.35	1.35

As in Table 3.16 the values/mile for both resistance and attenuation decrease as the gauge of conductor increases. The gauges shown in Table 3.16 are typical of British cable practice. French practice uses 0.4, 0.5, 0.6, 0.8mm.

Traditionally copper conductors have been used with paper core insulation and lead sheathing. These have been drawn into earthenware or concrete ducts or provided with steel tape or steel wire armouring for direct burial.

In recent years copper conductors have been used with polyethylene insulation and polyethylene sheathing extruded over an aluminium foil moisture barrier. These have been drawn into PVC ducts or provided with steel tape or steel wire armouring with a polyethylene sheath overall for direct burial. A figure of 8 configuration with an integral moulding onto a suspension wire is also used for aerial cable.

In former years much use was made of direct buried armoured cables in East Africa. These can still be found and providing they are not moved can still give good service. Cable in duct is much in use in towns and aerial cable is much in use in suburbs and rural area where it is pole mounted.

Lead covered paper core cable is still found. Conductors are traditionally twist jointed and each joint is covered with a paper sleeve. Splice covers are formed of lead sleeves which are plumbed to the cable sheath on each side.

In more recent years the conductors have been jointed by means of crimps applied with the aid of a manually operated compression tool. These pierce the insulation and obviate the need for stripping the conductor. Splice covers are provided by means of heat shrink sleeves.

In some cases the cable sheaths are impregnated with petroleum jelly to prevent the ingress of moisture. This is only of use in certain countries as in sandy countries the combination of petroleum jelly and blown sand makes jointing well nigh impossible.

The number of pair sizes in which cables are manufactured differ according to the country of manufacture. Britain manufactures sizes of 100, 200, 400, 600, 800, 1000, 1200, 1600, 2000 pairs. The USA manufacture sizes of 100, 200, 500, 1000, 1500, 2000.

France manufactures sizes of 112, 224, 448, 896, 1792 and 2688. All the cables are built up of quads. The French system is based on a unit core size of 7 and repeats this throughout the system. 7 pair and 14 pair distribution points are used.

The sizes are designed to fit in with each countries methods of cable planning. To attemp to mix two causes problems

For instance the use of traditional British cable plan ning using American cable sizes results in a lack of flexibility concerning choice of sizes and an increased amount of stumped pairs.

16.2.7   North American overhead plant

A very high proportion of North American local line plant is erected overhead, usually on poles erected on the margins of the roads, even in relatively densely populated urban areas. These poles are sometimes shared by as many as five different utilities; including high and low voltage power, street lighting, telephones and cable television. Safety rules are laid down in a convention based on the Bell-Edison agreement. The neutrals of LV power lines are compulsorily earthed in the USA. All exposed metalwork wuch as telephone cable supporting strands, power cable supporting brackets and so on are bonded together and earthed at regular intervals so that linemen working on the poles do not receive shocks when touching metal hardware. It is relatively uncommon to see a pole which is used exclusively for telephone wires. Distribution points are not mounted on poles as is common in British practice. Instead Ready Access Terminals are installed over the aerial cable and supported by the steel suspension strands.

In fact it is easily possible to tell whether North
American practices are in use in a country from the predominance of aerial line plant, the absence of joints on
poles and the ubiquitous North American Ready Access Terminal. The British Practice of having long loops of cable
and the joint on a telephone pole appears very strange to
North American line plant staff.

The Ready Access Terminal resembles a black plastic
box between 300mm and 600mm long installed over the cable
directly in line, which is supported by the steel suspension strands. It is equivalent to a removable cable
sheath and provides for some of the pairs in the cable to
be led via wire tails to a terminal block mounted at the
bottom of the Ready Access Terminal. Drop wires are connected to screw terminals on this block.

The Ready Access Terminal is usually installed a few
feet from a pole so that a lineman using climbing irons and
a safety belt can work on it, and so that poles can be replaced without affecting the joint. The pole can also readily be used for anchoring drop wires. Where access to a subscriber's premises does not fall opposite a pole but, say,
opposite the middle of the span, the drop wire can be
anchored to the messenger strands supporting the telephone
cable by a hook device with a clamp.

In theory Ready Access Terminals are installed at frequent intervals (typically adjacent to every pole in an
urban area) and one span of drop wire should suffice to
reach the subscriber's property. In practice this is seldom achieved. Several telephone cables of sizes up to
three inches in diameter and festoons of drop wire betwen
poles are typical of the North American pole route.

To British engineers this aerial joint user form of
construction presents a temporary appearance and seems vulnerable to storms and to collisions from motor vehicles.
In Britain it would also lay itself open to criticism by
environmentalists as it looks unsightly. Its origins lie
primarily in the history of many North American telephone
companies which often started as family businesses with
very little capital. Since the second world war the provision of power and telephone lines in rural areas of the USA
have been subsidised by the Rural Electrification Administration which has laid down codes of practice which accepted the techniques described above.

The adoption in the USA of overhead line plant techniques which also require much maintenance may seem surprising in a country where labour costs are high. This is explained by the fact that the labour force is at the same
time highly trained, strongly motivated and well equipped
with mechanical aids. One North American company quoted its
ratio of field staff to specialist motor vehicles as 1:1.6.
Private contractors are also extensively used in the USA.
For example, along some long pole routes there are pole
mounted cable pressurisation units every few miles. They
take their power from the power lines mounted on the same

poles. Being of overhead construction these power lines are
vulnerable and failures are common. There is therefore a
private contractor whose task it is to rush small petrol-
electric plants to the pressurisation units to keep them
powered during mains failures.

Articles in the North American press sometimes describe storms which brought down all the telephone cables and
describe with pride the speed with which all the lineplant
was re-erected. Advocacy of underground telephone cables
has become widespread in the USA relatively recently. High
labour costs have resulted in widespread use of mechanical
ditchers and moleplough tractors to install directly buried
cables underground in rural areas, especially on long runs
along roads having wide margins. Under pressure from environmentalists there is an increasing tendency to bury
cables in new housing areas. Direct burial in accordance
with the 'dedicated plant concept' is therefore increasing.
It is likely, however, to take many years before this
trend has a noticeable effect on the predominance of overhead aerial cable plant.

16.2.8   North American cables

North American underground primary distribution cables
at one time usually employed copper conductors, wrapped
paper insulation and a lead sheath. As lead sheathing became expensive there was a change to a sheathing technique
which caused the method widely employed in the food can-
ning industry. This produced a mild steel sheath, tinned
on the inside and having a soldered seam. The steel is fab-
ricated with a circumferential corrugation to provide flexibility, and a polyethylene sheath is extruded overall.
This type of cable is called 'Stalpeth'. Later there was
a tendency to adopt extruded paper pulp insulation rather
than wrapped paper.

In courses of time the British poly-laminate sheath
and cellular polyethylene insulation was copied in the USA.
All cables are usually of twin formation. Quad cables are
virtually unknown. A valuable feature is that the larger
cables are normally fitted at the factory with pulling
eyes. It is also common for the cable to be supplied from
the factory with its sheath marked every five feet with
the running length. This is a very useful feature for installers and storemen. The air spaced cables are normally
supplied pressurised on the drum. Metal drums, steel banded
battens and pressurised cables reduce the incidence of
problems caused by nailholes in the sheath.

16.2.9   Jointing in the North American network

Joints (splice closures) on underground cables at
first basically comprised lead sleeves dressed onto a special take adhering to the polyethylene sheath. Later, mechanical closures are used, comprising split galvanised cast

iron segments bolted together with a polyisobutylene gas-
ket. A self amalgamating polyisobutylene tape was also used
to seal the closure to the cable sheath. Cables and closures were continuously pressurised with compressed air. A
high air loss was accepted. For planning the capacity of
continuous dessicators a figure of 14 standard cubic feet
of air per day per sheath mile was typical. In some areas
an air pipeline system ws used to ensure a supply of air
at every few joints along the cable. Some companies adopted jelly filled cables, especially for the smaller sizes.
For these various kinds of closure methods, such as mechanical closures or heat-shrink sleeves, were adopted. For
aerial cables polyethylene sheathed self supporting or
lashed cables became common. Mid-span joints using an aluminium version of the mechanical split closure and polyisobutylene tape were used.

16.2.10   Flexibility in the North American network

    Flexibility has in the past been achieved in the North
American network by the use of the Ready Access Terminal
and by 'bridge tapping' or paralleling pairs and assigning
first the non-paralleled pairs (the 'preferential pair
count'). When the exchange pairs become inadequate there
is much costly rearrangement to be done. Ready Access Terminals have now been identified as a major source of
faults. The cross-connection cabinet called a 'service area
interface' has been introduced in the last 20 years.

16.2.11   Planning rules in the North American network

    Line planning rules have been mainly: plan to the res-
istance limit, do not bridge-tap more than 6 kilofeet, load
lines exceeding 18 kilofeet, or 15 kilofeet, depending on
whether the North American 'fine gauge' or Unigauge' techniques are the more appropriate.
    The practice of loading subscribers' lines (rather
than increasing gauge in the British and European style)
with 66 mH coils at 4.5 kilofeet intervals or 88 mH coils
at 6 kilofeet intervals had a logical foundation. As the
loss is proportional to $\sqrt{wCR}$ the effect of increasing
gauge is to reduce resistance but normally to increase mutual capacitance (see Table 3.16), unless at extra expense
special twin cables with thicker insulation are used, as
on junctions and on some long subscriber lines. To act on
the capacitative reactance by compensating inductive reactance was the common solution. Since the 'limiting line'
was a loaded cable pair the impedance to line of both the
exchange and the subscriber's equipment was designed to be
high, for minimum mismatch in the case of a loaded line.
In more recent years loop extenders (see Chapter 14) and
subscribers' amplifiers have been used.

16.2.12 Practices introduced to deal with a shortage of cable pairs

2-way shared service was introduced into the East African local line network at times when there was a shortage of cable pairs. This introcued many maintenance problems as linemen were continually getting the legs of the pairs crossed with resulting confusion, as both subscribers were then in parallel and both received all the calls for one subscriber and if both were in, both answered. Attempts were made to mark the DP terminals and the drop wires with coloured paints and tape but this had little effect in practice and the only ultimate solution was to lay more cable.

Line concentrators were introduced in an attempt to utilise the fact that most residential subscribers have a very low calling rate. These were most successful in suburban housing areas where all the lines connected to the concentrator were residential lines with low calling rates. Trouble occurred when there was a mixture of residential and business subscribers, as the business subscribers with their higher calling rate tended to utilise the available lines back to the local exchange all the time, so it was difficult for residential subscribers to make a call. As experience was gained with the use of line concentrators, in areas where business subscribers and residential subscribers were mixed the business subscribers were provided with DELs if at all possible, leaving just residential subscribers on the line concentrator.

16.2.13   Practices adopted in rural areas

In addition to the use of overhead open wire and underground cable pairs radio systems are used to connect subscribers in rural areas to their local telephone exchange. Such radio systems can take the form of:

a.   single channel VHF/UHF radio systems
b.   point-to-multipoint subscribers' radio systems.

The single channel VHF/UHF radio system is simply a means of providing service without the use of wires. Depending on the nature of the countryside and the availability of a line-of-sight radio path distances of up to 30 miles can be reached. The radio terminal nearest to the exchange is normally mounted at a site which is high compared with the area being served. The terminals possess equipment which copes with 2-wire/4-wire conversion, ringing, off-hook and so on as well as the radio equipment itself.

A number of different manufacturers make such equipment world wide although it tends to be rather an expensive way of providing service. In East Africa this was covered by the practice of requiring subscribers over a certain dis tance from the local exchange to provide a capital contribution towards the cost of the equipment.

Point-to-point subscribers' radio systems are manufactured in the UHF/SHF frequency bands. More and more they are making use of digital techniques and terminals are often intelligent. This is really the radio equivalent of the line concentrator as use is made of a small number of radio channels to provide service to a larger number of subscribers. The base station is placed at a suitable site to achieve good radio propagation as line-of-sight paths are required between the base station and the subscriber. The base station may be connected to the local exchange by a radio link or by landlines. The subscribers may each be equipped with their own outstation radio or may share a common outstation radio to which they are each connected by means of landlines.

Systems are manufactured which are modular in concept and which can be built up like building blocks to provide the flexibility needed for growth of the system. Systems vary in the number of subscribers fed from one base station. Figures of 60 to 100 are not uncommon. Depending on the choice of base station site subscribers may be provided with telephone service at distances of up to 50 to 60 miles from the local exchange.

The problem of maintenance needs to be considered as the calibre of technician needed to maintain such a system is quite different from the calibre of technician required to maintain a magneto telephone. Also there is the need to make the technician mobile. Very often the maintenance visit to the outstation simply results in the exchange of a faulty pcb for a good one, faulty pcbs being brought back to a central depit for repair. Depending on the facilities available in the country concerned it may even involve flying the faulty pcbs back to the manufacturer for repair on a regular maintenance exchange basis.

In choosing sites for base stations it is adviseable to consider all the assets that a Telecommunication Administration possesses in the area to be served, particularly existing radio stations. Very often microwave repeater stations are chosen as base station sites as circuits can be dropped and reinserved from the microwave radio system and transferred to the base station for use to communicate with the rural subscribers. The choice of the site as a microwave repeater site in the first place usually means it is a suitable site for radio propagation.

16.3 TELEX

It is the practice nowadays in many overseas countries to install a satellite earth station and a stored programme controlled telex exchange as one of their first priorities in order to provide telex communications for governmental and business subscribers within the country and to link these in turn to the international telex network.

Very often in small countries such as Lesotho and Botswana one such telex exchange suffices to cover the whole country. The whole of the telex network within the country can then be considered to be a local telecommunications network as telex subscribers outside the capital city are connected to the telex exchange as long line telex subscribers.

The type of equipment in use in these countries to s serve such subscribers veries. It may consist of:

a) S + DX equipment
b) Open wire carrier systems with integral VFT circuits
c) 24 circuit VFT equipment
d) 46 circuit TDM equipment

the first three have traditionally been routed over analogue audio carrier circuits. The last can be routed over an analogue or a digital audio carrier circuit.

16.3.1   S + DX equipment

This equipment tends to be utilised at the end of the telex circuit nearest the telex subscriber. It is an abbreviated form of speech plus duplex in which one or more duplex telex circuits are provided on an analogue audio speech circuit thereby reducing the bandwidth available for speech. Available systems vary as to the number of duplex telex circuits provided in this way. Typical amongst them are S + DX, S + 3DX, S + 7DX, S + 10DX. The telex circuits are normally provided at the top of the 4kHz analogue audio bandwidth and are separated from the speech by means of a filter. The telex circuits are separated by 120Hz from each other and utilise the CCITT recommended frequencies which are the same as those employed on 24 circuit VFT systems.

The use of such systems bring problems of their own as the speech circuit may be being used to provide a second speech circuit on an open wire pair to a remote telephone exchange whilst at the same time carrying long line telex subscriber circuits. Thus consideration has to be given to telephone transmission and signalling as well as to telex transmission and signalling. There may be conflicts. For example the telephone channel may need to employ a frequency of 3825Hz for outband signalling. In which case it will not be possible to use an S + DX system. More usually in such cases ring down signalling is employed and it is possible to use an S + DX system.

16.3.2 Open wire carrier systems with integral VFT channels

In East Africa and in Lesotho use is made of 3 circuit open wire carrier systems which include an integral group of 6 VFT channels within the baseband. As with the

S + DX systems the telex channels are spaced 120Hz apart and utilise CCITT recommended carrier frequencies.

Care needs to be taken when utilising such systems as they are essentially designed for two-wire use on open wire pairs on overhead trunk routes and the transmit channels may typically be situated at the high frequency end of the baseband and the receive channels at the low frequency end.

The frequencies chosen for the telex channels must be chosen from the CCITT recommended frequencies to fit in with this frequency plan and the transmit channels will be a different frequencies from the receive channels.

This may be compared with the 4-wire situation where the bearer is normally an analogue audio telephone circuit and the transmit and receive channels of the telex circuit normally possess the same carrier frequency.

This is particularly important when considering the number and type of space pcb cards to order for maintenance purposes.

16.3.3   24 circuit VFT equipment

These systems provide 24 telex circuits working over a normal analogue audio telephone channel. Nowadays they are frequency modulated each carrier varying by $\pm$ 30Hz to achieve mark and space signals either side of the carrier frequency. The carrier frequencies are provided in accordance with CCITT recommendations. Each carrier is separated from the next by 120Hz.

The equipment may be sub-equipped to provide 6, 12, 18 or 24 telex circuits.

16.3.4   46 circuit TDM equipment

Nowadays time division multiplex is being used more often to provide telex circuits. The cost per circuit is cheaper provided the country concerned has a sufficient number of telex subscribers to warrant their use. Where the number of telex subscribers are few, it is better to use the 24 circuit VFT equipment and if necessary to sub-equip the system with 6, 12 or 18 circuits.

ACKNOWLEDGEMENT

In writing this chapter I wish to acknowledge my debt to my predecessor Paul Skey.

I found that I could not hope to improve on his sections on North American practices, which he clearly knew intimately, so I have included these sections almost untouched from the previous edition of the book.

Chapter 17

Future evolution

Ian Dufour

17.1 INTRODUCTION

The local network, sometimes known as the local-line network, the access network or the local-loop, is the vital link from the customer's premises to the telephone exchange and thence to the junction, trunk and international networks. For any telephone company it represents a very substantial proportion of its assets and as all customer services ultimately rely upon this part of the network, the management and development of it is of fundamental importance.

Hitherto, the network has been based on analogue transmission over metallic-pair cables and the majority of existing plant assets reflect this comprising predominantly outside plant items such as duct (underground conduits), manholes and joint-boxes, cables, cabinets and poles. Today, however, there are substantial influences at work which will result in significant changes occurring over the coming years. These influences are that:-

- There is demand for digital services led by business customers

- Optical technology is vastly increasing the capabilities available to the network planner and providing cost-reductions at the same time.

- Switching and transmission technologies are converging around common digital "building blocks". This in-turn blurs some of the conventional network distinctions between switching and transmission and between local and junction networks.

- Software, for providing flexibility, control and monitoring, is beginning to dominate hardware in the complexity of new systems.

- The complexity of new systems means that the system components are increasingly provided only by major international companies to meet a world demand.

- Technological developments and changes to regulatory frameworks are permitting telephone companies to be "by-passed" by competitors.

- The long-term integration of telephony and Cable-TV is widely foreseen around the world, albeit on an indeterminate timescale.

The changes resulting from these influences can be summarised by saying that local networks will evolve from being predominantly analogue circuits based on metallic-pairs towards a network having significant proportions of digital circuits employing a substantial element of electronics. The demand for this change will be led initially by business customers but in the long term may well be dominated by residential demand for visual and inter-active services. Apart from the need to develop suitable system components this evolution has implications for planning strategies and the network structure itself as well as for planning, installation and maintenance practices. Some of the effects of this evolution will be examined in this chapter.

17.2 CIRCUIT REQUIREMENTS FOR DIGITAL SERVICES

Until recently, data transmission requirements have mostly fallen into the 0-9.6kbit/s range and these can comfortably be accommodated on existing audio local plant using modems located in customer's premises. However, many developments are in-hand which will extend the data-rate requirements. By far the most important of these is the development of digital PBXs which, together with the trends towards an Integrated Services Digital Network (ISDN) is leading to the availability of a range of digital terminal equipment (for example, computing workstations, facsimile, slow-scan television) operating at 64kbit/s. The availability of this terminal equipment will lead to a demand for digital transmission both at 64kbit/s and also at 2Mbit/s to link a PABX to a digital exchange so that terminals can have access to the digital junction, main and international networks and hence to other users.

To some extent the availability of digital terminal equipment before the public switched network is capable of handling it is also leading to an increase in requirements for digital private services; this is being met by British Telecom (BT) with the **KiloStream**, **MegaStream** and **Packet SwitchStream** services, which also have their own clearly separate long-term markets.

The **KiloStream** service offers user rates of 2.4, 4.8, 9.6 or 48kbit/s with control and supervisory facilities, or 64kbit/s without. The control and supervisory facilities are provided via a 6+2 coding structure giving rates of 3.2, 6.4, 12.8 or 64kbit/s. The three lowest rates are sent to line at 12.8kbit/s (the two lowest being reiterated to form a 12.8kbit/s signal), and the remaining options at 64kbit/s. The 12.8 or 64kbit/s signal can be transmitted to line over a 4-wire circuit using the encoding technique known as WAL2, but the preferred method is now to use a variation of an echo-cancelling digital 1+1 system to provide service over a 2-wire circuit.

The ISDN, which has been described in detail in Chapter 12, provides customers with speech and/or digitally presented services based on a 64kbit/s switched network. The BT junction and trunk transmission networks and the trunk and local exchanges are already committed to a digital conversion programme (figure 1a) which will be largely complete by the end of the 1980's but the provision of full ISDN rests upon digital transmission being taken directly to the customer; that is, **integrated digital access (IDA)** (figure 1b). These services are marketed by BT as **single-line IDA** and, for the 2Mbit/s service for digital PBXs, **multi-line IDA**.

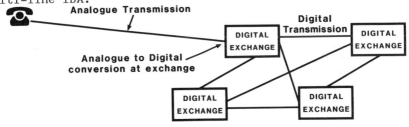

Fig. 17.1a   Digital Main and Junction Networks

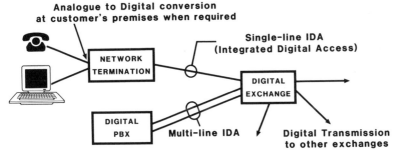

Fig. 17.1b   Integrated Services Digital Network (ISDN)

Single-line IDA, which provides two user channels, transmits information over an existing pair in the local network.   It terminates at the customer's premises on network terminating equipment (NTE), and the user then has access to a range of standard interfaces for the connection of terminals.   These can cover CCITT X and V series terminals at transmission rates of up to 64kbit/s as well as high-quality telephony terminals.

However, it is the provision of 2Mbit/s services such as **Multi-line IDA** and **MegaStream** that represent a watershed for local network development because, whilst the 64kbit/s services just described can if required be served over existing plant, in general 2Mbit/s services cannot.   It is this fact which leads to the necessity for the provision of new cables and transmission equipment and by doing so it causes a complete re-appraisal of the way the network is structured and developed.

17.3 FUTURE TRENDS IN SERVICES

The services described above allow a wide range of digital facilities, both public-switched and private, to be provided now. However, this is still very much the beginning of the digital era in the local-line network. Much more is technically possible and, whilst the market justification for a great deal of it is still unclear, several trends are already apparent.

Rapid Introduction of Digital PBXs

The attractions of digital PBXs for integrating office computing and communications will become rapidly apparent to users, and a fast rate of introduction is likely as more designs enter the market and existing PBX's reach the end of their useful lives. The rate of change could well be encouraged by PBX designs using local area network (LAN) ring structures and related techniques for internal communication. The added advantages that then accrue from their connection to a digitally-switched network giving internationally recognised and compatible ISDN facilities will further increase the pace of change. One effect of this will be to make 2Mbit/s provision to business customers a commonplace event. Something which, as we will see later on, will in turn have a considerable effect on the design of the local network.

Cable-TV Systems

In many countries Cable-TV systems have existed for some time, but until relatively recently they have been thought of as only supplying domestic consumers with entertainment channels. Now there is a prospect of providing telephony and related interactive services to residential customers, and also digital services to business customers as well as vision services. The impact of Cable-TV on the design and development of the local network could become very significant but the effects are more likely to be quite long-term and strongly influenced by market demand as well as by regulatory influences. Nevertheless, the dream of a "fibre to every home" will ultimately rely on the commercial viability of supplying vision channels combined with a variety of other services. Whilst this commercial impetus is presently weak in the UK for existing services there are reasons why this could change. The trend towards integration of telecommunications and Cable-TV technology is generally quite well recognised throughout the world and considerable research effort is being directed towards its achievement. A possible factor leading towards implementation of such systems will be a demand for High-Definition Television (HDTV) channels which cannot reasonably be supplied in multi-channel form from terrestrial transmitters and which will require a wide-bandwidth medium for their delivery. This could be by satellite broadcasts in the microwave spectrum but is more likely to be provided by an optical-fibre cable system.

Metropolitan Area Networks

The concepts, techniques and developments applicable to LANs for within-building data-distribution can easily be extended to inter-building and even inter-community applications where they are known as Metropolitan Area Networks (MANs). In many cases the systems use Cable-TV types of cable and their associated electronics technology. Whether or not MANs are integrated with consumer vision services they provide an interesting addition to conventional systems and offer challenges in network design as the compatibility of the data-ring approach with the conventional star network requires careful evaluation.

17.4 DIGITAL PLANT

As stated earlier, 2Mbit/s represents a watershed in the plant requirements for digital services. Below this bit-rate, existing cable pairs are capable of supporting the 64kbit/s services by using up to 144kbit/s transmission systems. Above 144kbit/s there is a twilight zone where perhaps 6 channel, 384kbit/s, and 10 channel, 704kbit/s, systems may have some limited application but, in general, the higher the bit-rate the greater the problems with crosstalk and attenuation become. In the local network there is also an enormous variety of signal types with potential for mutual interference. This, coupled with a regular re-arrangement of circuits due to new and ceased customer orders, means that the continuing management of high reliability high bit-rate digital circuits is increasingly difficult with the procedures long-since put in place to cater for an analogue network. Although some countries feel that their existing cables can support some 1.5 or 2Mbit/s services this is not the case in the UK and special plant provision for 2Mbit/s services is required. This can take the form of:-

- a new metallic-pair cable purpose designed for 2Mbit/s digital transmission (transverse-screen cable) together with 2Mbit/s line transmission systems similar to those used regularly in the junction network.

- a new optical-fibre cable with associated line transmission systems

- microwave radio systems

Optical fibres will be dealt with in more detail in a later section in this chapter but it is useful to provide a summary of the other 2 options at this stage.

Transverse-screen Cable

Transverse-screen cable (see also Chapter 14), whereby the two directions of digital transmission are separated by a diametric aluminium foil screen to control near-end crosstalk, has been used extensively in the UK junction network and in the local network to provide 2Mbit/s service to **MegaStream** customers. It has the advantage of being low-cost and simple to install using existing procedures and skills but its disadvantages are that regenerators are required on circuits over 2km long (and at every 2km thereafter) and that 2Mbit/s is the effective upper limit of operation.

Microwave Radio Systems

Microwave radio systems subdivide into point-to-point and multi-point systems and examples of each are illustrated by local-network applications in BT.

Point-to-Point

At present, in BT, customer links using 19GHz point-to-point equipment are provided in some circumstances. They are most useful for expedient relief purposes for services at 2Mbit/s and higher, typically to provide service quickly until a new cable can be provided. Systems operating at 29GHz have also been evaluated and are in limited use and further purchases are being considered. Both these systems are suitable for 2 and 8Mbit/s services, with possible future variants catering for higher bit-rate and video applications.

Multi-point

A 19GHz multi-point Radio system is also available in BT for taking low bit-rate (for example, 12.8, 64kbit/s) circuits to customers that are difficult to serve by more conventional means. The system also has a capability for delivering 80kbit/s, 144kbit/s and even 2Mbit/s. A central station serves a sector that is typically 90° or 120° wide and covers up to a 10km distance. Within this sector, outstations receive and transmit low bit-rate channels selected from an aggregate bit-rate at the central station of about 8Mbit/s, and typically about 90 x 64kbit/s channels can be served in one sector.

General

In the UK radio systems are not widely used in the local network and this is likely to remain the case in the medium-term, although developments in low-cost millimetre-wave technology could change this situation significantly in the longer-term. The present limited use of microwave radio systems is partly because in city and town centres, where the major business customers tend to be, it is frequently impossible to get "line-of-sight" links and partly because the existing infrastructure of duct makes cable systems more cost-effective. However, in other parts of the world where cities are predominantly low-rise but geographically widespread the use of radio is common. Radio is also a key technology for competitors to "bypass" established telecommunications providers.

17.5 OPTICAL FIBRE CABLES AND SYSTEMS

As already indicated optical-fibre cables are the other alternative for the provision of customer service. Where the customer requires service above 2Mbit/s (ie 8Mbit/s, 34Mbit/s or 140Mbit/s) then the transverse-screen cable option is not technically viable and, as has been seen, radio systems are a minority solution. In such cases optical-fibres have for some time been the only practical solution as indeed they have for special applications such as providing service to power-stations where electrical induction inhibits the use of copper pairs. The cables and line terminals used in these applications have essentially been multi-mode junction-type hardware, re-engineered on an ad-hoc basis for customer links. It is, however, the use of optical-fibres for 2Mbit/s provision where the challenges lie for the provision of optical-fibre to become widespread before the commercial requirement for visual services.

It is a commonplace understanding in transmission engineering that wide-bandwidth multiplexed systems rely on long route distances and large circuit demands to justify their provision. The local network is the opposite of this case and it might be wondered why optical-fibres should have a place at all. There are several reasons why they should. The first is that the costs of optical technology are reducing more quickly than competing technologies. Furthermore, as will be seen later, the potentials of optical technology are still enormous, notwithstanding the astonishing developments so far, and new capabilities and further developments can be expected for many years ahead. The second is that the conventional network in which there are high capacity links in the trunk network, smaller capacity links in the junction network and a pair in the local network for every service offering can be, in effect, "turned on its head." By using an optical-fibre link to a customer requiring a variety of services the link can act as a high capacity "pipeline" to carry all the various services (figure 2). Initially this will apply to business customers where analogue exchange lines, **single-line IDA**, **multi-line IDA**, analogue leased-lines, **KiloStream**, **MegaStream** and other services can share a common optical "pipeline" but eventually the concept can apply to residential customers so that one or two exchange lines, inter-active services and

vision channels can be routed over their own optical
"pipeline". A final reason is that the transmission system
design, if chosen carefully, can permit much longer route-
distances than is conventional in local networks and this
can lead to economies in the placement of exchanges within
the network.

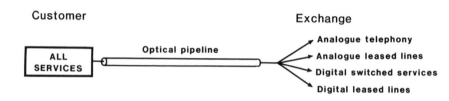

Customer

Exchange

Optical pipeline

| ALL SERVICES |

Analogue telephony
Analogue leased lines
Digital switched services
Digital leased lines

Fig. 17.2 The Optical Pipeline.

17.6 THE OPTICAL PROMISE

The major optical developments in probable
chronological order can be summarised as:-

- the emergence of single-mode fibres as the world
  standard. This is already an unstoppable trend. There
  are fundamental reasons of manufacturing processes and
  material costs why single-mode fibres will be
  consistently cheaper than telecommunication-standard
  multi-mode fibres. The alternative approach of using
  large-diameter plastics fibres inhibits the ease of
  service provision and network re-structuring that results
  from a common fibre type throughout the trunk, junction
  and local networks.

- reduced cost optical devices (transmitter/receivers).
  Hitherto, low-cost devices (LED/PINFET) have been
  restricted to 850nm low bit-rate (2Mbit/s and 8Mbit/s)
  operation over multi-mode fibres. The difference in cost
  between a transmitter/receiver pair of this type and a
  high bit-rate (140 or 565Mbit/s) pair for 1300nm trunk
  applications (LASER/APD) currently exceeds a ratio of
  100:1. The use of single-mode fibres has until recently
  been associated with the higher cost devices. However,
  considerable work is being carried out into the use of
  light-emitting diode (LED) and light edge-emitting diodes
  (ELEDs) over single-mode fibres, as well as the design of
  low-cost lasers. A range of devices at reasonable cost
  is already emerging.

- controlled wavelength transmitters. In light output
  terms an LED can be compared to "white noise" whereas a
  laser is essentially a spot frequency device. The
  problem is that the wavelength (ie frequency) of

operation cannot yet be closely determined at reasonable cost. When it can, and it can be contained to a narrow linewidth, the prospect of several, if not many, channels operating within a common optical bandwidth "window" - say 1300nm+25nm - is opened up. Thus between 1275nm and 1325nm there could be several channels at 10nm intervals each having a capacity of hundreds of Mbit/s.

- availability of splitters (couplers). At their simplest a splitter will cause a single light source to be sub-divided into 2 equal paths. This can have applications in broadcasting a common signal - as with Cable TV - or in conjunction with different wavelength light sources over the common input channels with the separate wavelengths being detected on each of the two outputs. Splitters with more than 2 outputs can also be designed giving rise to a number of other applications.

- cost-effective wavelength multiplexing. This will depend on the devices and components described above. Wavelength multiplexing between 2 different wavelength windows (say 1300nm and 1550nm) is likely to be practical before multiplexing in the same window. In general, the cost-effectiveness of optical wavelength multiplexing against conventional time division multiplexing from high capacity systems will depend on a complex interaction of system component costs.

- development of optical switching. This is at an early stage but R&D work is in progress around the world and will undoubtedly find application.

    The net effect of these is that a network of cables containing single-mode fibres will have an expansion capability for many years to come. The potential bandwidth available will permit services to be provided that are not at present thought viable, and that may include HDTV and a range of other unforeseen services. In a decade's time technology will be able to deliver the component parts for a ½Gbit/s switched network, although whether it will be commercially justified and hence materialise is another matter.

17.7 SWITCHING AND TRANSMISSION TECHNOLOGIES MERGE

    Whilst one vision of the future is the use of optical technology for both transmission and switching a similar trend with the existing, more conventional digital technology, is already apparent today. Digital transmission systems now use 2Mbit/s as a building block but so also do digital exchanges. Inter-exchange links already connect to the exchange at the 2Mbit/s level and on the customer side provision is being made in exchange designs for **multi-line IDA** connections at the 2Mbit/s level. At the component level the integrated circuits and design principles are already common. But it is at the system level that interesting dilemmas occur.

As digital transmission is made available in the local network the possibility arises of placing multiplexed systems out in the network. At its simplest this would involve a multiplexer interfacing at an analogue level into an exchange at one end. At the other end of a digital link there would be another multiplex mounted in a building, a cabinet or an underground structure interfacing with the copper local distribution network. This configuration is already common in North America where local-line lengths are long and much of the growth is on the periphery of towns and cities. A development of this is that when the exchange is a digital type the multiplex at the exchange end can be omitted with consequent savings in system costs. Whilst the multiplexers at each end can be designed as stand-alone items the integration of one end into the exchange itself calls for strong system inter-relationships. The same result can be achieved by physically relocating the "front-end" of the exchange in the network and treating it as a multiplexer. The situation is further complicated when the remote multiplexor also has concentration facilities (ie more customer lines are connected than the capacity between the concentrator and the exchange) because there are then similarities with the design of small stand-alone exchanges and this raises a further range of system design considerations.

For switched network services working to a digital exchange through remote electronics connected to the exchange at the 2Mbit/s level there is virtually only a semantic difference between a multiplexer and the "front-end of the   exchange". However, when leased-line services are also considered a new set of considerations arise in that the leased-line channels must be "stripped away" at the exchange end and sent to other networks. There are several ways of handling this, one of which is to use the exchange itself but there are others using an additional and separate 64kbit/s switching stage (called a Service Access Switch or, in North America, a Digital Cross-connect) in conjunction with the exchange. The location and application of the Service Access Switch (SAS) within the network is also complex and an emerging science in its own right. The networking implications are further complicated when switching at bit-rates of 2Mbit/s and higher are considered. An SAS capable of handling 2Mbit/s is likely to be available before exchange switches are configured to cater for it but in the longer-term it can be expected that exchange designs will also cater for switching higher bit-rates than 64kbit/s.

The inter-relationships of transmission and switching systems are increasingly complex and system designers are faced with a variety of options many of which are still to be fully understood in their impact on the costs and management implications of running a network. In spite of this it can confidently be predicted that there is an international trend amongst telecommunications operators towards fewer, more complex, exchanges with remote customer interface units located within the local network at large

customer sites and perhaps in cabinets or underground
structures, and linked by digital signals over single-mode
fibres. The local network will thus also cover larger
geographic areas and increase in average length of circuit
and in fact "take-over" parts of what was previously called
the junction network.

17.8 SOFTWARE

As many of the trends already outlined above become
implemented there will be a very distributed network of
inter-related electronics. The interfacing of all these
systems will be complex and equal to some of the most
complex computer architecture design problems. Furthermore,
the need for centralised control, performance monitoring,
fault monitoring and maintenance control will put increasing
emphasis onto "network management" capabilities which in
turn will be based on computers. The net effect will be
that a significant proportion of the cost of system
development will be in software - a trend already well
established with digital exchanges. The limited
availability throughout the world of appropriate software
skills and the difficulty of managing such large projects
could well provide constraints on the pace of developments.
Certainly, few manufacturers and telephone companies will be
capable, individually, of designing a fully inter-related
digital switching and transmission network with clearly
defined operating and management procedures and capable of
co-existing with the old analogue systems. The developments
in the local network will only serve to make this point more
strongly.

17.9 CABLE-TV

Reference has already been made to the probable
convergence of conventional "telephony" services and Cable-
TV. So far, commercially viable Cable-TV systems around the
world generally rely on co-axial cable technology although
there have been numerous trials using optical fibres in at
least some part of the system. Whilst there is no sign of
this position changing in the near-future it is, as has been
indicated earlier, widely regarded internationally that one
day optical fibres will be used in this application as well.
There is thus at least a prospect of "a fibre to every
home". Some of the constraints on this happening, and the
commercial and technical solutions adopted to allow it to
happen, will vary from country to country depending on the
regulatory framework within which telephone companies and
Cable-TV companies operate.

17.10   BYPASS

Just as the new digital service demands and the
technology options present telephone companies with a range
of opportunities to radically change their own networks so
there are opportunities for newcomers. Where the regulatory
framework permits it telephone companies can compete for the

provision of service to customers. Access to the customer
might "bypass" the traditional local network completely suc
as by satellite or terrestrial radio links, duplicate it by
the provision of additional and separately routed cables,
probably optical, or interconnect with it at some higher
level in the network such as at a main switching centre. I
countries where the legislation does not permit "bypass" th
technological trends will increasingly provide pressures
from customers for relaxations in current practice.
Interestingly, the new technologies which permit "bypass" c
the traditional copper-pair local network are equally
available to the established operators and the end result i
more likely to be a convergence of technical design rather
than an evolution of fundamentally different facilities.

17.11 CONCLUSION

The preceeding paragraphs have outlined some of the
changes occurring in the coming years which will affect the
local network. It is the last part of most countries
networks to be modernised and some of the driving forces
have been described. In such a large and costly part of th
network complete change will be a slow process and the
copper pair will remain in use for decades yet.
Nevertheless, the advent of electronic systems located at
customer sites and in cabinets, and linked to exchanges (if
not actually considered part of them) by optical systems
will provide significant changes in the planning and
management of the network. The changes will be so
significant that they will also flow back into the network
and have fundamental effects on exchange design, the nature
of their support systems, and the structure of a total
national telecommunications network of exchanges and
transmission systems.

Chapter 18
ISO seven layer model for open systems interconnection

Ron Brewster

What is normally referred to as a "Telecommunication System" is, in reality, a communications infrastructure in combination with a set of service applications. In order to understand all of the requirements placed upon a system, it is necessary to make a proper distinction between the communications functions and the applications that they support. It must also be recognised that the communications functions are not concerned solely with issues of the physical and network connnections, but that the transport, session control and presentation of the information carried are of equal importance. This recognition must be accompanied by a rational view of the address structure required, the differences between application identity (or high level address) and the network and physical addresses, and the realisation that directory services are needed to give flexibility to the users and not to items of equipment. This approach is the basis of the 7-layer model for communications systems developed by the International Standards Organisation (ISO). The arrangement is shown in the following diagram.

Layer	Level		Level
Application	7	◄------►	7
Presentation	6	◄------►	6
Session	5	◄------►	5
Transport	4	◄------►	4
Network	3	◄------►	3
Link	2	◄------►	2
Physical	1	◄------►	1
	Physical connection		

The ISO 7-level model for open systems interconnection.

The logical and physical partitioning of the functions of a complete communications system is conveniently realised with such a layered concept in mind. Although the concept is an abstract one and the implementation need not necessarily have its physical partitions bonded to the layers of the model, the value of the concept in the clarification of the design issues involved cannot be underestimated.

Dr.Ron Brewster is with the Department of Electrical and Electronic Engineering and Applied /Physics, Aston University, Birmingham.

The concept of Open Systems Interconnection (OSI) is that data terminals may be interconnected by the network in an unrestricted way. The ISO model identifies seven levels or layers for the definition of the various protocols and interfaces necessary for Open Systems Interconnection. At the lowest level, level 1, physical parameters for signals are defined simply to enable signals to be transferred over the physical connection. This layer, the Physical Layer, consists of specifications for the line code, transmission rate, signal voltage levels, physical connectors and other parameters to enable satisfactory transfer of streams of digits over a simple single connection path. At this level, no significance is attached to any single digit, it is concerned only with the efficient transfer of data across the physical connection.

The second layer, the Link Layer, is concerned with the establishment of a disciplined and reliable data link across the physical link. At the physical level there is an inherent degree of unreliability in that there is no knowing whether errors have occured in the transfer of data. Thus one of the major functions of the Link Level is to provide error detecting and/or correcting facilities. This in itself may require the division of the data stream into blocks so as to identify the field over which any error-detecting code operates. Alternatively, block delimiters may be necessary for synchronisation purposes. Although various digits now perform specific functions, nevertheless the basic transparency of the data link is maintained in the bits allocated to the data field in the Link protocol.

A data network comprises a number of nodes interconnected by various data links. The third layer in the OSI model is concerned with network operation and is known as the Network Layer. At this level addresses are required to specify the links to be used to interconnect the appropriate terminal nodes. This may specify logical links rather than specific physical paths, since the same physical path may not necessarily be used for successive packets of the same message transaction. Control and acknowledgement fields are also provided to ensure that data communication is correctly established between the appropriate network nodes. The network control should be independent both of the Link Level of control and the higher order levels of protocol.

The three layers so far considered are concerned only with the communications network and are thus properly the concern of the communications engineer. The fourth layer, the Transport Layer, takes into account the nature of the terminal equipment and is concerned with establishing a transport service suited to the needs of this equipment. It must thus select a link through the network which operates at a data rate and quality appropriate to the needs of the terminals involved in the communication operation, thus relieving the user from concerning himself with the detail of the mechanism of data transfer through the network.

We are thus beginning to depart from the basic task of providing a communication path through the network.

The next three layers, layers 5 to 7, are task-oriented and have to do with the operations performed by the data terminal equipment rather than with the network. The Session Layer is concerned with setting up and maintaining an operational session between terminals. It can thus be basically identified with the operation of 'signing on' the computer to begin the operation of the desired task and 'signing off' to signify the completion of the task.

The Presentation Layer is concerned with the format in which data is to be presented to the terminals and resolves differences in representation of information used by the application task. Each task can thus communicate without knowing the representation of information (e.g. data code) used by a different task. The purpose of this layer is to make the network machine-independent.

The Application Layer defines the nature of the task to be performed.   It provides the actual user information processing function and programs for application processes in the real world, for example, airline booking, banking, electronic mail, word-processing etc.   These three higher order layers are mainly concerned with the organisation of the terminal software and are not directly the concern of the communications engineer.   The Transport Layer is the layer which links the communication processes to these software-oriented protocols.

The basic philosophy of the 7-layer model is that each layer may be defined independently of every other layer.   Thus from the user point of view, interchange takes effect across each layer as shown in the broken line connections in the diagram.   In fact each operation passes down through the layers of the model until data interchange is effected through the physical connection.

There are numerous examples of the application of the OSI model to specific systems described in the literature.   The following books contain representative examples:
1. R.L.Brewster (Ed) "Data Communications and Networks", Peter Peregrinus 1986.
2. R.L.Brewster "Telecommunications Technology", Ellis-Horwood 1986.
3. K.G.Beauchamp "Computer Communications" Van Nostrand Reinhold 1987.

Chapter 19
Flexible network architectures
David G. Fisher

19.1 What more do you want?

We have seen how ISDN will bring the benefits of digital
telecommunications into the local loop. What more is needed?
ISDN basic access allows 144 kbit/s capability over existing
wire copper pairs and primary rate access at 2 Mbit/s over 4
wire circuits usually requiring new cables.   While improved
techniques may stretch these capabilities a little, a complete
recabling with optical fibre will be required to extend the
capacity to be able to handle broadband services. Such a
venture will require extensive capital investment. New
architectural concepts are needed to minimise these costs and
to use the new resources to the greatest effect particularly
during the introduction phase.

In this process of evolution of the local network we can
predict a number of trends:

- further integration of service access, on a digital
  basis, to include both communicative and distributive
  broadband services;

- distribution of intelligence to allow more effective
  control of resources;

- user control to provide network resources on-demand.

Recognising these trends and the driving forces for change, a
target network architecture is described and an evolutionary
process established.

19.2 The Forces for Change

- User Needs   There is a growing understanding and
                recognition of end user needs. The
                increasing influence of powerful business
                customers, for whom communications is of
                strategic importance, will drive network
                operators away from the traditional,
                regulated utility approach to service.

David Fisher is with STC Telecommunications, New Southgate.

- Regulatory Environment
Globalisation of telecommunications and gradual deregulation means that all network operators are subject to competition at some level.

- Political and Economic
There is a better understanding of the power of telecommunications, making it the subject of regional, national and international policy. Simple public service obligations have been supplanted by considerations of the economic value of telecommunications services.

- Technology.
There are two key enabling technologies which impact the local network; VLSI and optical. The former allows the introduction of electronics to better exploit the potential of existing cables while optical fibres will be the basis for a fundamental upgrading.

19.3 Broadband Services

Extension of ISDN bandwidth capability allows new services to be introduced. The main new opportunity is for high quality video communication. Larger bandwidth also improves the quality potential for existing services. Fig. 1 shows the relationship between services and bandwidth.

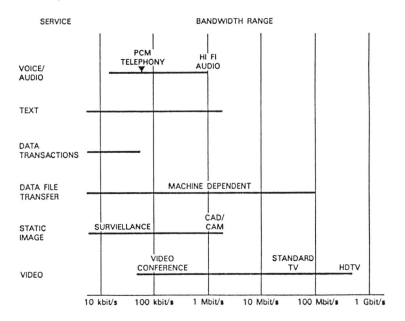

FIGURE 1
TELECOMMUNICATIONS BANDWIDTH REQUIREMENTS

Each type of service, audio, data or video can be provided ove
a large range of bandwidths. Within the range the value of
bandwidth provided by the network has two types of impact;
quality and cost.

For audio services, for example, telephony quality voice
transmission using simple coding techniques requires 64
Kbit/s. This has been the basis of today's ISDN. More comple
coding schemes, such as ADPCM, allow either a reduction of
bandwidth demand to 32 kbit/s or an improved quality at the
same bit-rate. Even lower bit-rates can be used if slightly
lower quality is acceptable. Because the frequency band
available for cellular mobile telephony is a limiting factor i
is planned to use 16 kbit/s coding for the digital version. C
the other hand broadband ISDN will allow High Fidelity stereo
music transmission at rates around 1 Mbit/s. In the case of
data services, the network bandwidth capability determines the
length of time needed to transfer a particular quantity of
information.

As shown in Fig. 1, the range of bandwidths required for video
communications is very large. If we consider today's broadcas
quality colour TV encoded into digital form, using relatively
simple techniques, then the bandwidth required would be of the
order of 150 Mbit/s. There is however a large amount of
redundancy in the information. This is both in space (large
areas the same colour) and in time (parts of the picture which
do not change from one frame to the next). Sophisticated
video-codecs can take advantage of this redundancy and reduce
the transmitted bandwidth to around 30 Mbit/s without
unacceptable loss to quality. Alternatively the same
techniques can be used to improve to the levels required for
large screen displays; this is normally referred to as High
Definition TV. In the special case of video-telephony or
video-conferencing, it is expected that higher bandwidth
compression will be built into the video-codecs. This is
firstly because of the higher level of redundancy in the
information to be transmitted and also because it involves lor
distance two-way transmission rather than broadcasting from
relatively close "head ends".

The overall conclusion is that a great variety of bandwidths
should be accommodated in a Broadband ISDN. Moreover
individual services will shift their optimum bandwidth demand
as technology advances and higher quality can be economically
provided to the user. Hence the need for a new,
bandwidth-transparent transfer mode: ATM which will be
explained later.

Services have been considered above in terms of their differen
bandwidth requirements. Other aspects are also important in
determining network requirements. CCITT have defined the
following classification:

- Interactive Communications Services

  • Conversational services, e.g. telephony, video-conference

  • Messaging services, e.g. electronic mail

  • Retrieval services, e.g. video library

- Distribution Services

  • Without user control, e.g. broadcast TV

  • With user control, e.g. remote education.

Having looked at service requirements let us now turn to the design of local networks that will handle them.

19.4 Network Architectural Concepts

The starting point is a "Reference Configuration". This shows a number of "Functional Groupings" separated by "Reference Points". Standardised interfaces may be established at these reference points. Fig. 2 shows a reference configuration for the local area segment of an ISDN.

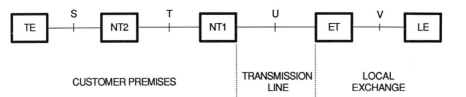

CUSTOMER PREMISES   TRANSMISSION   LOCAL
                              LINE   EXCHANGE

FIGURE 2
ISDN REFERENCE CONFIGURATION

TE -   Terminal Equipment

NT2 -   Network Termination 2 - provides terminal
        distribution on the customer premises network

NT1 -   Network Termination 1 terminates the transmission
        line at the customer's premises

ET -   Exchange Termination - terminates the transmission
        line at the local exchange

LE -   Local Exchange - provides connection related function

Interfaces may be provided at the Reference Points of Fig. 2 as
follows:

S - terminal interface

T   -   user/network interface (CCITT)

U   -   user/network interface (USA)

V   -   exchange line interface.

Chapter 12 has explained the implementation of these interfaces
for the British Telecom ISDN pilot. CCITT has produced
international standards for basic rate (144 kbit/s) and primary
rate (2 Mbit/s [Europe] or 1.5 Mbit/s [USA]) user/network
interfaces. It has also generated an outline of the
architecture for a target Broadband ISDN (rec. I.121). This
recommendation envisages Broadband user/network interfaces at
150 Mbit/s and 600 Mbit/s.

The general network architecture outlined above can be expanded
to show the details of different arrangements in the means of
connecting customer's premises networks to local exchanges and
other public network facilities. At this stage we must
recognise two alternative basic approaches; the first is the
"distributed exchange" concept and the second is "managed
transmission networks".

The distributed exchange concept is shown in Fig. 3.

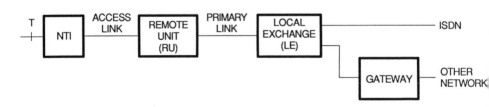

FIGURE 3
DISTRIBUTED EXCHANGE CONCEPT

Remote Units (RU) have been added between the customer's premises and the parent local exchange. These can be either multiplexors or concentrators. In both cases they serve to separate the local transmission links into Access Links which are provided per customer and Primary Links which serve several customers. This enables a more economical introduction of new services and new technology into the local area by:

- reducing the total amount of cable required

- allowing cable upgrading in two stages (primary followed by access)

- extending the potential serving area of the local exchange.

The second approach provides the same benefits in a different way as shown on Fig. 4.

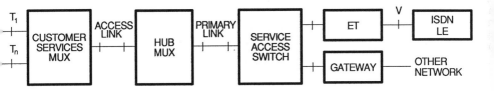

FIGURE 4
MANAGED TRANSMISSION NETWORK CONCEPT

The basic difference in approach is that the Managed Transmission Network concept recognises the existence of several logical telecommunications networks sharing a common transmission network resource while the distributed exchange considers that all services will be provided by or accessed through a universal ISDN.

The Managed Transmission Network concept involves a "Customer Services Multiplexer (CSM) which combines channels carrying different types of services onto the access link. The Hub Multiplexor (HM) combines several access links onto a smaller set of primary links. The Service Access Switch (SAS) is a digital cross-connect device providing separation of service channels to the appropriate network facility. The configuration of the multiplexing and cross-connect allocations in CSM, HM and SAS are all controlled by a Network Manager.

The rest of this chapter describes the implementation of the Managed Transmission Network approach and the way it is expected to evolve with advancing technology.

19.5 The British Telecom City Fibre Networks

This is a good example of a Managed Transmission Network using today's technology. Put into service in 1986, its objective is to use optical fibres to serve the needs of large businesses and customers with similar needs in the City of London. The system is shown schematically in Fig. 5

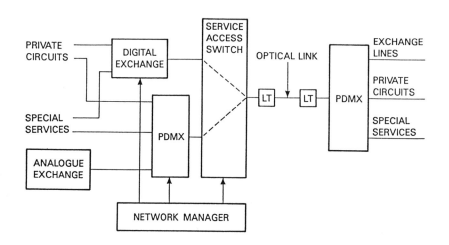

FIGURE 5
BT CITY FIBRE NETWORK

Comparing this diagram with the general architecture of Fig. 4 we can see that the CSM function is performed by the PDMX which is STC's flexible, Primary Digital Multiplexer. The Service Access Switch is implemented by GPT using System X technology. The Hub Multiplex function is not shown but is located at both ends of the optical transmission link in the form of conventional Higher Order, Plesiochronous Multiplexers.

Probably the most important element in this network is the network manager, which provides control of the service access switch and PDMX, to enable managed point to point connection through such a network.

The PDMX has a bus structured backplane allowing the insertion of a variety of interface cards for subscriber access. For example, for analogue voice, digital voice, data interfaces and so on. The PDMX can be over-equipped with more subscriber access cards than are needed at any one time, so that the ixture of interfaces available to the customer can be changed to meet his changing pattern of needs. For example, predominantly voice circuits during the day can be changed over at night to data circuits, to download information between computers. From the network operator's point of view, he can manage the transmission infrastructure to maximise his revenue by controlling the service access switch and the multiplexers. He can set up point to point connections, through a network management console, such as this when they are needed and with the required services interfaces. So if a particular circuit, let's say an X25 interface, is required from one location to another between 3 o'clock and 5 o'clock on Friday afternoon, that can be set up under the control of the network manager. The circuit can be broken down again when it's not required, and the equipment in the network reconfigured to provide access for someone else.

The benefits arising from this network approach are realised through controlled reconfiguration providing service variety. There are benefits to both the user, who will have to pay less for the service access he requires, and to the network operator because he can earn more revenue from the same capital investment in equipment.

19.6   The Role of Synchronous Multiplexers

An emerging standard for a world-wide digital synchronous hiaerarchy is likely to give rise to a new generation of higher order multiplexing equipment which will have the following characteristics.

- make efficient use of high rate digital transmission on optical fibre

- embedded opto-electronic interface

- drop/insert capability

- grooming capability

- digital cross-connection at higher bit-rates.

The basic principle of synchronous multiplexing is that there is precise positioning of the individual channels at rates of 64 kbit/s or above within a high rate (say 150 Mbit/s) bit stream. Individual channels can then be inserted or extracted without having to decompose the full structure. The "grooming" facility involves repositioning channels within the time frame. This is equivalent to time switching. To understand the way in which grooming is used we must look at the structure of the digital synchronous hierarchy in more detail.

A set of draft CCITT recommendations have been prepared based
largely on work done by Bellcore in the USA on an arrangement
they call SONET (Synchronous Optical Network).   Fig. 6 shows
the CCITT draft standard structure for a bit-stream at 155 520
kbit/s using a 125 microsecond frame. The frame is drawn as a
rectangle with 270 bytes per row and 9 rows per column.   Each
byte corresponds to a 64 kbit/s channel.

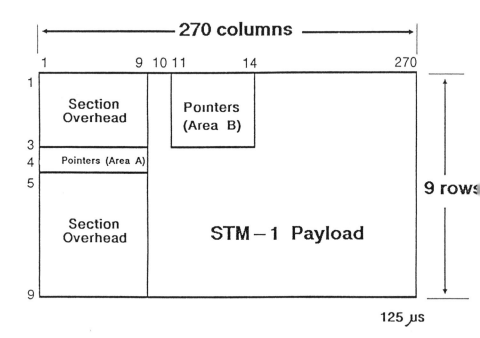

FIGURE 6
DIGITAL SYNCHRONOUS STRUCTURE AT 156Mbit/s (STM - 1)

The significance of the rectangular presentation is that blocks
of adjacent columns can be treated as entities and handled more
easily by the electronic implementation.   These blocks are
called "Adminstrative Units" in CCITT and "Virtual Tributaries"
in SONET.   The grooming facility is then used to assemble
relevant channels into Administrative Units for
dropping/inserting by a synchronous multiplexer or handling
within a synchronously organised cross-connect.   Fig. 7 shows
the application of these techniques within a managed local area

transmission network. Signalling Adjuncts are shown on the
diagram. These are required whenever common channel signalling
is employed. They handle the routing of signalling information
to correspond with the grooming of user channels.

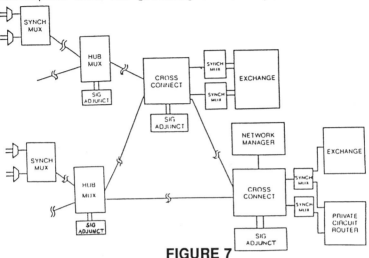

FIGURE 7
APPLICATION OF SYNCHRONOUS MULTIPLEXING

19.7 Asynchronous Transfer Mode (ATM)

ATM is a form of time division multiplexing and switching in
which all information is packetised for insertion in time slots
which are identified by a label rather than their position in a
synchronised frame.

Some of the implications of the ATM technique compared with
conventional time division are:-

- ATM provides inherently statistical multiplexing and thus
  has built-in traffic concentration;

- Each call can use as much or as little of the available
  bandwidth, by taking as many timeslots as it needs;

- Services with different bandwidth requirements can be
  multiplexed together in a dynamically flexible fashion.

- The technique is fundamentally more efficient as
  resources are only used when information is to be
  transmitted. There is no need to transmit silence!

Because of these characteristics we can predict that ATM will
in due course replace conventional techniques for both
multiplexing and switching. The key question is how? Clearly
we must look at those sections of the network which would

benefit most from the bandwidth transparency of ATM.   One such
area is flexible network access for business subscribers as
shown in Fig. 8.   This scheme shows where an ATM subsystem
(hatched) would connect subscribers into a conventional
existing network (plain).

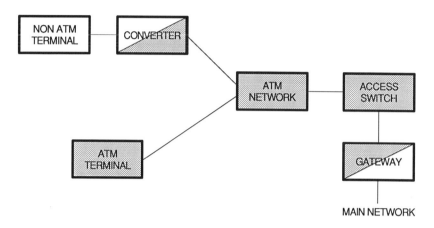

FIGURE 8
INTRODUCTION OF ATM

19.8   Optical Techniques

The next major development in networks is likely to be optical
signal processing.   Here we are once again seeking to use not
just the enormous potential bandwidth of the optical carrier
but the ability to realise flexible access to it.

The frequency of light is two hundred thousand Gigahertz.   If
we could exploit light, as we do radio waves at frequencies ten
thousand times lower, then we can have the modulation
flexibility we desire.

An analogy of today's optical systems being like the use of
spark gap transmitters in the early days of radio would be
quite accurate.   We should therefore see how we can take
advantage of that very high frequency carrier by using radio
type techniques.

Fig. 9 illustrates the application of these radio-like techniques to build optical networks where we can connect process controls, data, telephone or videophone onto a single fibre with light being fed from the two ends for bi-directional transmission.

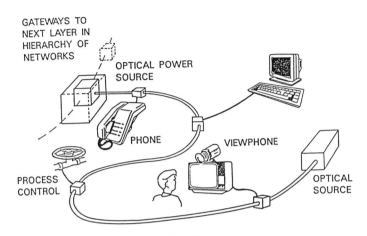

FIGURE 9
NETWORK WITH OPTICAL SIGNAL PROCESSING

19.9 How will it all happen?

The changes we have discussed in local telecommunications network will involve very large investments in product development by equipment suppliers and in the purchase and deployment of equipment and cable by the Network Operators. Political initiatives are needed to create the environment in which such investments can be made with adequate expectation of reasonable returns. At the European level the Commission has to take a number of steps towards the creation of such an environment. One of these is the establishment of the RACE (Research and development in Advanced Communications technologies for Europe). This is a five year sponsored collaborative R&D project targetting the creation of an embryonic Integrated Broadband Communications network across Europe by 1995. It involves all the major European network operators and telecommunications equipment manufacturers as well as other interested parties such as potential users and service providers. It is organised in three main parts:

I deals with overall network architecture, introduction strategy and potential usage.

II is the development of appropriate technologies for video terminals, video codecs, optical transmission, ATM and optical switching, etc.

III introduces verification facilities and application pilots.

19.10 References

1.   CCITT Recommendations of the series I, Integrated Services Digital Network (ISDN). Red Book, Vol III, Fascicle III.5, 1984.

2.   IEEE Journal on Selected Areas in Communications - Special Issue on Digital Cross-Connect Systems. January 1987 Volume SAC-5 Number 1.

3.   Thomas A., Coudreuse J.P., Servel M. "Asynchronous Time Division Techniques: An Experimental Packet Network Integrating Video Communication". ISS '84 Proceedings. Vol. 3 paper 32C2.

4.   Gallagher I.D. "Multi-service Networks". British Telecom Technology, Vol. 4 No. 1 Jan. 1986.

5.   Radley P.E., Fisher D.G. "Evolution of Subscriber Access Networks", ITU Forum 87, Geneva, October 1987.

Chapter 20
Optical fibre networks
John O'Reilly

20.1 Introduction

Optical fibre communications has developed rapidly over the last twenty years: from little more than a gleam in the eye of two imaginative researchers in the U.K. [1] - closely parallelled by near synchronous studies in France [2] - to the dominant medium for digital point-to-point telecommunications links at rates ranging from 2Mbaud to 565Mbaud and higher. The sheer ubiquity of the technology, encompassing a wide range of signalling speeds, network topologies and applications - makes this far more than just another means of achieving signal transmission: it represents a new era in telecommunications in which the synthesis of optical fibre transmission with photonic switching seems set to change beyond recognition the face of the telecommunications network. And while optical fibres have to date been deployed primarily in the trunk and junction networks, this too is changing. Local network applications are seen as a very major growth area for optical fibre systems.

In this chapter the basic principles of optical fibre transmission are reviewed briefly. This is followed by consideration of a number of systems applications with an emphasis on evolving network topologies and possible implications for local telecommunications. No attempt is made to be encyclopaedic in coverage; this would be quite impracticable in a work of this scale. Rather we seek to present examples which highlight special features which can be expected to prove of importance in several guises and in a variety of applications.

20.2 Optical Fibres and Their Characteristics

There are two broad classes of optical fibre, multimode and singlemode, both of which find applications in telecommunications. Considering first the case of multimode fibres, here the central light-carrying core is relatively large, compared with the wavelength of light, so that propagation may realistically be described in terms of ray optics. Multimode fibres may be of *step index* or *graded index* form. With the former the refractive index is uniform within the central core region stepping abruptly down to some lower value in the surrounding cladding while graded index fibre has a gradual reduction in index from a maximum value at the centre of the core down to a lower value at the cladding. Light travels in a step index fibre by a series of total internal reflections at the core-cladding interface and in a graded index fibre by a process of continuous refraction, as shown in Figure 1.

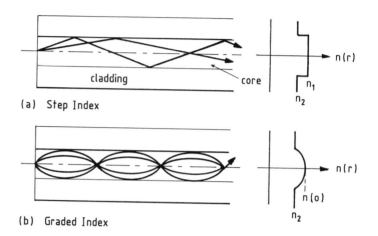

(a) Step Index

(b) Graded Index

Figure 1.   Ray paths and refractive index profiles
for multimode fibres

In both cases there are multiple paths by which light can travel along the fibre from source to destination so that if these paths have different propagation delays then pulse broadening will occur.   This is clearly the case for step index fibre. The pulse spreading is proportional to length, L, and also to the refractive index difference $\Delta n = (n_1 - n_2)$.   Denoting by ΔT the temporal spreading we have for the step index case:

$$\Delta T = \frac{Ln_1 \Delta n}{cn_2}$$

Pulse spreading limits the distance-bit-rate product which can be achieved so that a small value of Δn is desirable from this point of view.   Unfortunately small Δn limits the acceptance angle for the fibre and exacerbates coupling and jointing problems.

For the graded index fibre while there are multiple paths these all have nearly the same transit time.   Pulse spreading is very much reduced compared with a step index fibre and this made possible the operation of first generation 140 Mbit/s optical fibre systems over distances of 10km-15km.

These first generation systems operate at optical wavelengths close to 850nm where fibre loss is, by modern standards, rather high - 3 to 4dB/km.   At the longer wavelengths of 1300nm and 1550nm the loss is much reduced, as illustrated in Figure 2, so that much greater transmission distances can be contemplated.   One consequence of this is that even with graded index multimode fibre the multipath pulse distortion can become significant.   This leads to a consideration of

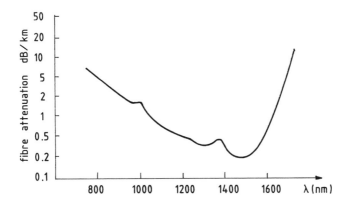

Figure 2. Attenuation versus wavelength for a silica-based fibre

singlemode fibre. Typically a multimode telecommunications fibre has
a cladding diameter of 125μm and a core diameter of 50μm. In
contrast, a singlemode fibre while still having a cladding diameter of
125μm has a core only a few μm in diameter. Consequently propagation
cannot realistically be described in terms of ray optics. It will
suffice here to say simply that a single mode fibre provides
essentially a single path from source to destination, overcoming the
multipath limitations of step index and graded index multimode fibre.

That is not to say, though, that pulse spreading or *dispersion*
does not occur with singlemode fibre - it does, but is caused by a
different mechanism, namely *chromatic dispersion*. In general
refractive index is a function of wavelength so if an optical pulse
comprises a range of wavelengths - due to finite source linewidth -
these will experience different delays and the pulse will spread out.
In addition, the waveguide structure of a monomode fibre results in a
wavelength-dependent delay. It is thus the combination of these two
components - material dispersion and waveguide dispersion - which
determines the total chromatic dispersion characteristic of a
singlemode fibre. Chromatic dispersion may be quantified in terms of
pulse spreading per unit of source linewidth per unit of transmission
distance: [ps nm^{-1}km^{-1}]. Considering just the material component we
have for silica fibre approximately 80 ps nm^{-1}km^{-1} for λ=850nm, near
zero for λ=1300nm and 18 ps nm^{-1}km^{-1} for λ=1550nm. It is possible
using special fibre designs to effect a degree of cancellation between
waveguide and material contributions to total chromatic dispersion so
that singlemode fibre can be produced with essentially zero dispersion
at the very low loss wavelength of 1550nm [3]. The increased cost and
complexity of these special fibre designs has to date militated
against their use for telecommunications applications and so-called
standard singlemode fibre is generally used with total chromatic
dispersion close to the values given above for silica.

For local network applications where distances are relatively short and bit rates are relatively low graded index multimode fibre is entirely suitable.   However, it is now more expensive to manufacture this than standard singlemode fibre so there is potential benefit in deploying singlemode fibre in the local loop.   The main disadvantage is that a laser source is required to achieve good coupling to a singlemode fibre.   As we shall see later, though, this consideration is not necessarily paramount when put alongside the observation that laser sources are considerably more expensive than simple light emitting diodes.

20.3   Optical Sources

Both laser diodes (LDs) and light emitting diodes (LEDs) are used for optical fibre systems.   Laser diodes have considerably narrower emission spectra (perhaps ~1nm) than do LEDs (~20nm-40nm).   For these and other reasons lasers are the preferred choice for long haul high bit rate applications.   For short haul modest data rate applications, though, LEDs are generally preferred.   Direct intensity modulation of LDs and LEDs is achieved by variation of the drive current. Illustrative optical power versus current characteristics are shown in Figure 3.

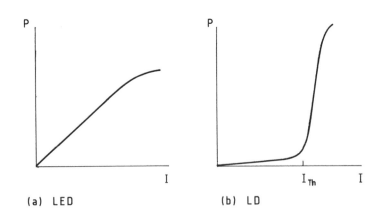

(a)   LED   (b)   LD

Figure 3.   Output optical power versus drive current for
(a) light-emitting diode and (b) laser diode

The power available from a laser diode is of the order of 1mW ($\equiv$ -0dBm) and for an LED typically an order of magnitude less, say 0.1mW ($\equiv$ -10dBm).   These levels are representative of the power launched into a multimode fibre.   For coupling a laser to a singlemode fibre we might expect to achieve approximately -3dBm launched power while the mismatch between an LED and a singlemode fibre is so great that the coupling loss for this combination is tens of decibels. Nevertheless, this combination has recently attracted attention for short distance applications since fibre path loss is then so low as to make acceptable this high coupling loss [4,5,6].   A particular attraction of this strategy is that it preserves the option of subsequently upgrading the system by changing the terminal equipment.

20.4 Receivers for Optical Fibre Communications

Optical detection is readily accomplished using a semiconductor photodiode. Incident photons are absorbed within the device releasing electrical charge carriers to take part in conduction. The detected optical signal thus manifests itself as a photo-induced current which may be passed to electronic amplification circuitry. Early, short wavelength, systems use silicon photodiodes.. These are often designed such that photon detection initiates within the device an avalanche multiplication process to provide low noise internal gain. These avalanche photodiodes (APDs) provide sensitive receivers but require high bias voltages, in the region of 100V to 200V.   In contrast, receivers based on silicon PIN photodiodes involve only a slight degradation in performance and offer the convenience of 5V to 10V operation. For longer wavelengths silicon is not a suitable detector material and devices have been developed based on the GaInAsP quaternary system.   Receiver sensitivities, corresponding to satisfactory operation for digital transmission, are of the order of 1000 to 2000 detected photons per bit. This may be compared with a limiting value of approximately 11 photons per bit, which is related to the quantum nature of light.   Clearly there is some scope for improvement here, with practical receivers operating some 15dB to 20dB away from the quantum limit. Figure 4 provides an indication of receiver sensitivity as a function of bit rate.

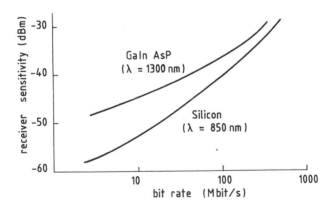

Figure 4.   Illustrative variation of receiver sensitivity
with bit rate

It should be stressed that the receivers being discussed at this point employ *direct detection*, essentially making no use of photon phase and frequency information. We shall see shortly that *coherent detection* can offer improved receiver sensitivity - approaching the quantum limit - and allow a greater variety of system/network topologies.

20.5 Direct Detection Systems

Present day installed systems rely on direct detection. Given a transmitter power of the order of 0dBm and a receiver sensitivity for 1300nm operation varying with bit rate from about -50dBm to -30dBm this allows a loss range of 30dB to 50dB. On the basis of a cabled fibre loss of 0.5-1.0dB/km this indicates an unrepeated transmission range of the order of 30km to 100km, depending on bit rate and actual fibre loss. These figures are generally realistic but make no allowance for joints and repair losses, dispersion penalty, operating margins, etc. To allow for these factors rather lower repeater spacing targets were adopted by British Telecom in 1984 for their Standard Optical Fibre Systems, the main technical features of which are summarised in Table 1.

Application	8/34Mbit/s Graded Index	140Mbit/s Graded Index	140Mbit/s Single Mode
	Junction	Trunk	Trunk
Repeater Spacing	10-15km	10km	30km
Maximum System length	30km	630km	630km
Wavelength	1300nm	1300nm	1300nm
Source	LED	LED/Laser	Laser
Receiver	PIN-FET	PIN-FET	PIN-FET
Fibre Dimensions	50/125μm	50/125μm	9/125μm
Fibre attenuation	<2dB/km	<2dB/km	<1dB/km

Table 1. Main Technical Features for BT Standard Optical Fibre Systems (1984)

The unrepeatered spans of Table 1 are, of course, considerably greater than is required for local network applications so that there is considerable flexibility available here.

20.6 Direct Detection Local Network Systems

The use of optical fibres in the trunk and junction networks is now well established but for the local network there have to date been only a very small number of installations due to the cost advantage of metalic cable systems. This position is now changing rapidly. There has recently been a concerted effort to reduce the cost of optical fibre systems suitable for the local loop to facilitate significant penetration in this area. Trials in several countries have been conducted on imaginitive solutions combining singlemode fibre technology with low cost LED transmitters [4,5,6]. One such scheme [7] is illustrated in Figure 5. Notice that this makes use of standard singlemode fibre technology designed for 1300nm and 1550nm operation. This fibre is not, in fact, singlemode at the operating wavelength of 820nm, this wavelength having been selected because of the availability of low cost LEDs and silicon PIN receivers. Even

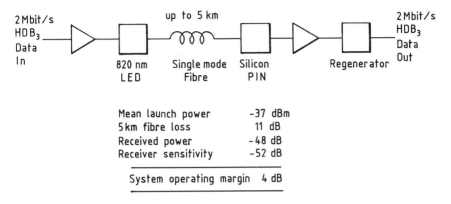

Figure 5.   Low cost 2Mbit/s optical fibre system [7]

with the very high coupling loss associated with an LED source and the relatively high fibre loss associated with operation at λ=820nm system spans in excess of 5km are obtained which is quite adequate for local loop applications.   Note also that, unusually for optical fibre systems, HDB₃ has been adopted rather than binary signalling.   This is because HDB₃ is a standard interface code.   The slight degradation in receiver sensitivity introduced by use of HDB₃ signalling is more than outweighed by the cost savings achieved by eliminating conversion between HDB₃ and binary signal formats in the terminals.

If significant use of optical fibres is to be achieved in the local network and at customer premises then low cost installation techniques are required.   A very significant step in this direction is provided by *blown fibre* technology [8].   This technique involves the installation of empty plastic tubes, formed as a cable using traditional methods.   Subsequently, packaged optical fibre bundles are blown into the tubes using compressed air.   Fibre may be blown in over distances of at least 500m and tandem blowing techniques enable greater lengths to be installed.   This arrangement is particularly attractive for provisioning customer premises such as large commercial buildings.   The plastic tube 'ducts' may be preinstalled with fibre being deployed at a later date in response to demand.

In deploying optical fibres into the local network we are not at present, of course, seeking to provide 2Mbit/s replacement for individual subscriber lines.   Rather, fibre technology is used to extend high bit rate time division multiplexed transmission into the local network, further towards the subscriber than was formerly the case.   With this arrangement, multiplexing and concentrating equipment is located at 'Cabinet' and 'Distribution Point (DP)' level and fed via fibre links from the local exchange.   This is one small part of a trend which is beginning to blur the distinction between transmission systems *per se* and the multiplexing and switching functions traditionally located within the switching centres.   This is a trend which is set to continue apace and be given further emphasis under the accelerating influence of coherent optical technology.

20.7   Coherent Detection

Up to this point we have concentrated on systems based on direct detection.   There is considerable effort, though, being directed towards coherent systems using optical heterodyne and homodyne detection.   This relies on combining the incoming optical signal with an optical local oscillator wave in such a way that coherent mixing takes place at the photodetector.   This occurs because a photodiode is essentially a square-law device so far as the incident electric field is concerned.   The operation may be appreciated as follows:

Let the incident signal be given by

$$E_i(t) = m(t)\cos\omega_c t$$

where $m(t)$ represents the information bearing message waveform and ω_c is the angular frequency of the optical carrier.   The local oscillator signal we shall denote as:

$$E_{LO}(t) = A\cos\omega_{LO}t$$

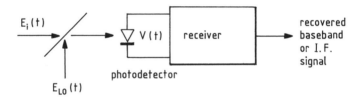

Figure 6.   Schematic representation for coherent detection

The detection arrangement is shown in Figure 6 and the output from the (square-law) photodetector has the form

$$V(t) \equiv \overline{(m(t)\cos\omega_c t + A\cos\omega_{LO}t)^2}$$

$$\equiv \overline{m^2(t)\cos^2\omega_c t + A^2\cos^2\omega_{LO}t + 2m(t)A\cos\omega_c t\cos\omega_{LO}t}$$

Here the overbar simply indicates that $V(t)$ does not contain terms in the region of twice the input optical frequency, so that it may be written as:

$$V(t) \equiv 1/2m^2(t) + 1/2 \ A^2 + m(t) \ A \cos \ (\omega_c - \omega_{LO})t$$

For A, the local oscillator signal level, sufficiently large the first term may be neglected while the second term represents a d.c. component which carries no information. The essential output signal is thus:

$$V(t) = m(t). \ A. \ \cos \ (\omega_c - \omega_{LO})t.$$

If $\omega_{LO} = \omega_c$ this is a baseband signal corresponding to homodyne detection. For $\omega_c \neq \omega_{LO}$ we have heterodyne detection with $(\omega_c - \omega_{LO})$ representing the (angular) intermediate frequency and $m(t)$ is ultimately recovered by a second detection process just as in a superheterodyne radio receiver. Notice that for homodyne detection phase synchronism is required between the optical local oscillator signal and the incoming optical signal. This requires an optical phase locked loop while in contrast, for heterodyne detection, frequency tracking will suffice. There is, though, a slight performance advantage for homodyne detection and both techniques are investigated in research laboratories.

The primary motivation for coherent optical detection is that it offers the prospect of much improved receiver sensitivity. The conversion gain of the optical mixing process depends on the local oscillator injection level and if this is sufficiently high then near quantum limited detection is achievable: an improvement over direct detection of perhaps 15dB. The major price we pay for this improvement is in receiver and transmitter complexity – including the need for semiconductor lasers with high spectral purity. A further advantage, though, is that coherent detection makes possible fine-grain wavelength division multiplexing (WDM) – essentially optical frequency division multiplexing – which opens up the possibility of novel and very flexible optical network architectures.

20.8 New Network Options

Wavelength division multiplexing may be implemented on a coarse-grained basis using sources with wavelengths sufficiently widely separated that they may be individually selected by optical filters. If the latter are electrically turnable then we may select at a given receiver any one of a multiplicity of optical signals – each with separate wavelengths – propagating in an optical 'ether'. Use of WDM in combination with coherent optical technology, though, greatly increases the capabilities of this general strategy [9] providing in principle for entire optical networks of large dimension with signal routing and selection being effected optically. An illustrative example of such a network structure is presented in Figure 7. For convenience the transmitters are shown on the left and the receivers on the right but in practice the network is folded such that these are co-located in pairs. Coherent optical detection enables a receiver to be 'tuned' to select the signal from a given transmitter terminal. Alternatively, fixed-tuned receivers with optically tunable transmitters enable the transmit terminal to selectively route a message to a given destination. Notice that while a receiver tunes in

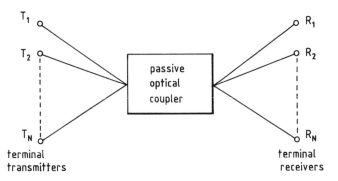

Figure 7. Passive coupler-based optical network

to select a particular wavelength the optical power is nonetheless divided across all potential receivers. The sensitivity improvement offered by coherent detection is of benefit here in allowing the signal to be more finely divided.

Given a suitable network topology it is possible to group wavelengths into 'bands' to facilitate distribution and also to limit the power division indicated above.

We may note here that the number of channels potentially available to use by the combination of singlemode fibre technology and coherent detection is very large indeed. With a fibre loss of <1dB/km there is in the region of 30,000 GHz of optical spectrum available in each of the low loss windows near 1300nm and 1550nm. The scope for imaginitive use of this potential spectrum is considerable and it should certainly provide ample opportunity for the development of optical local telecommunications networks. The question for the future, perhaps, is just what do we mean by local? With optical wavelength division routing and other photonic switching schemes providing for the migration out into the network of what have traditionally been exchange concentration and switching functions it is likely in the future to be difficult to 'see the join'.

20.9 Concluding Remarks

This chapter has sought to provide a brief overview of optical fibre principles and techniques in so far as they impact telecommunication networks, with particular attention given to the likely future impact on local telecommunications. Oddly, perhaps, our general conclusion is that while optical fibre transmission has to date made a very significant entry into the trunk and junction networks it seems poised to have yet more wide-ranging impact on the local network. The combination of singlemode fibre and photonic switching technologies offer the potential to largely eradicate the local exchange - or perhaps we should say distribute its functions within the network, leading to an optical transmission and switching continuum.

References

1. Kao, K.C. and Hockham, G.A., 'Dielectric-fibre surface waveguides for optical frequencies', Proc. IEE, 113, 1151-1158, 1966.

2. Werts, A., 'Propagation de la lumiere coherente dans les fibres optiques', L'Onde Electrique, 46, 967-980, 1966.

3. Imoto, et al., 'Characteristics of dispersion free single-mode fiber in the 1.5µm wavelength region', IEEE J. Quantum. Electron. QE-16, 1052-8, 1980.

4. Cochrane, P., et al., 'A solution to low cost local optical transmission on optical fibre', Proc. IEEE Conf. Lasers and Electro-Optics, CLEO-86, June 1986, 252-3.

5. Kaiser, P., 'Single-mode fibre technology for the subscriber loop', Tech. Dig. IOOC-ECOC, Venice, October 1985, Vol.2, 125-8.

6. Krumpholz, O., 'Subscriber links using singlemode fibres and LEDs', ibid. 133-9.

7. Cochrane, P. and Reeve, M.H., 'A low cost optical line transmission system', Proc. Sino-British joint meeting on optical fibre communications, Beijing, P.R.C., May 1986, 51-54.

8. Hornung, S., et al., 'The blown fibre cable', IEEE J. Selected Areas in Commun., August 1986, SAC-4, No.5, p.679.

9. Stanley, I.W., et al., 'The application of coherent optical techniques to wideband networks', IEEE Jnl. Lightwave Technol. LT-5, April 1987, pp.439-451.

Index